Pollen Microscopy

by

Norman Chapman

Pollen Microscopy

by Norman Chapman

Second Edition

Edited by Valerie Rhenius
Pollen drawings by Norman Chapman
Photographs by Norman Chapman and Valerie Rhenius
Sue Carter
Gerry Collins
Sally Dunn
Bill Fisher
Janet Morris
Stephen Rhenius

Stock photos www.istockphoto.com: Ash male flower in bud, small balsam, christmas cactus, water crowfoot, harebell, cross leaved heather, honesty, horseshoe vetch, *Limnanthes douglasii,* rock rose, germander speedwell, stonecrop, strawberry tree, snow in summer, sun spurge, and wood sorrel.

Copyright © 2018 CMI Publishing Ltd
Individual photographs © Norman Chapman, Valerie Rhenius or the photographers listed above

All rights reserved.
No part of this publication may be reproduced, stored in or introduced into
a retrieval system, or transmitted in any form or by any means
(electronic, mechanical, photocopying, recording, scanning or otherwise)
without the prior written permission of the copyright owners.

ISBN: 978-1-907092-11-4

Published by
CMI Publishing Ltd
Baythorne Cottage, Baythorne End, Halstead, Essex, CO9 4AB, UK
www.cmipublishing.co.uk

Printed and bound by Ashford Colour Press Ltd, Gosport, Hants PO13 0FW   www.ashfordps.co.uk

Contents

	Page
Foreword	1
Pollen Studies	3
Pollen, Wind, Insect pollination	3
Cross Pollination	3
Pollen morphology	4
Form	5
Collecting Pollen	5
Direct from the flowers	10
'Touch on' to the microslide	10
More difficult pollen	10
Various pollen forms under the microscope	6
Transparency	6
Polyads	6
Tetrahedrons	6
Furrows	6
Unique characteristics	7
Hairs	7
Jelly	7
Family characteristics and exceptions	8
The exine	8
Difficult identifications	9
Useful kit	11

	Page
Microscopy	12
Microscope power	12
Stereo microscopes	12
Coupling camera to microscope	13
Setting up a compound microscope	13
Preparing the pollen grain for microscopy	14
Preparing Slides	14
Getting the pollen on to the slide	14
Mountants	16
Making the drawings	17
Bees and Pollen	18
Pollen from a trap on a honey bee hive	19
Stray Pollen	21
Extracting pollen from honey for microscopy	23
Pollen in honey	24
Drawings: Pollen grains and their source flowers	25
Acknowledgements	257
References	257
Index	259

Foreword

Norman Chapman kept several hives of honey bees in Surrey, south east England for many years. In addition to his inventiveness regarding the gadgetry of beekeeping which resulted in his published book "Constructive Beekeeping", he developed a keen interest in the pollen collected by his bees.

This furthered an interest in microscopy, and led to Norman joining the Quekett Microscopical Club, to which in 2017 he was elected an Honorary Member, for promoting microscopy and the Quekett, and for his willingness to share his equipment and techniques for making slides of pollen.

Over a period of several years Norman began collecting pollen samples, viewing them under a microscope, and making detailed drawings of the pollen grains from garden flowers, native wild flowers and trees.

The first edition of "Pollen Microscopy", published in 2015 included about 100 of Dad's drawings. Since then he has added a few more, and we've added more photos, aided enthusiastically by collegaues.

An even greater selection of these drawings, many chosen for their interesting source, shape, form, or from plants whose nectar and pollen is of importance to the insect world is presented in this, the second edition of Dad's book.

Valerie Rhenius
(Norman's daughter)
2018

Pollen studies

The exine, the tough outer skin of pollen, can survive for many years, long after the remainder of the pollen grain has dissolved and been lost. Pollen identification can give information on the plant origin, geographical, seasonal and historic source of the pollen, which is useful in forensic as well as archaeological study.

Because honey bee individuals forage on one species at a time, studying the pollen they have collected gives important information about where and when they visited the pollen source. Beekeepers can derive useful information on the day to day foraging of their bees by studying the colours of the pollen pellets collected in a pollen trap. Pollen analysis can also be used to identify the source of honey and wax.

Pollen

Most plants, that is all plants that produce seeds, produce pollen. Pollen contains the male genetic material, which, when combined with that from the female part of the flower, enables viable seeds to be formed. Cross pollination allows more diversity in a species, and the variety can be important for survival.

Plants have developed two main methods for delivering pollen grains from the original plant to another of the same species.

The wind is used to disperse pollen by many plants eg grasses, conifers and catkin bearing trees. This needs the production of large quantities of pollen of a size to be carried by the wind. These grains tend to be within the range of 20 to 40 micrometres (µm). They are too small to see with the naked eye. Thirty placed end to end would measure about a millimeter (mm). They are blown around, then when the wind abates, these pollen grains fall towards earth. If they fall within range of the female stigma, the grains are drawn towards the receptor by electrostatic attraction. Tons of these grains fall everywhere each year, sometimes as visible dust. The hard outer skin, the exine, persists in the environment yielding interesting forensic information upon analysis.

Some wind borne pollens are commonly responsible for the allergic reaction known as hay fever.

Insect pollination, carried out mostly by bees of all species, is used by many plants. Pollen grains from insect pollinated plants range more widely in size than those carried by wind: 6µm to 200µm, most being around 30µm. Flowering plants and bees have co-evolved over around 130 million years to their mutual benefit and survival.

There is a trade off with insect-pollinated plants offering food (pollen and nectar) in exchange for transference of some of the pollen to neighbouring flowers. Smaller quantities of pollen are needed than for wind pollination.

Services to plants by honey bees are particularly effective through their habit of constancy ie visiting the same flower species sequentially, offering great economy in cross pollination. After depositing the small amount needed for pollination, it has been estimated that an average colony of honey bees will collect 44 kilograms of pollen in a season, most of which goes into the hive to feed the larvae.

Cross pollination. Plants have evolved ingenious devices to favour cross pollination and avoid self pollination. For example, some, like white bryony, holly, willow and yew, are dioecious, bearing male and female flowers on separate plants. Others, for example hazel, birch and marrow are monoecious, having unisexual flowers but both sexes on the same plant. Most flowers are bisexual with the male and female parts maturing at different times to avoid self pollination. Among the grasses the sex of a flower can vary depending upon the position of the individual spikelet on the spike.

Pollen morphology

Pollen grains vary enormously in colour, size, shape and form. The more so in pollen of insect vectored plants. Most are near the average of 30μm, and approach the spherical.

Size: the most accurate way to measure grain size is to use a stage microscope, a glass slide which has a series of fine parallel lines scored on the surface, in conjunction with a linear graticule, set within an eyepiece.

Viewed under that objective which is used to inspect grains, a factor can be derived to give a figure for any grain sized under that objective.

An alternative is to compare size against the reference of hazel pollen, noted for being reliably just under 25 microns.

Shape: the spherical appearance of pollen gains can be deceptive. Some are oblate, slightly squashed from the poles like the earth. They sit flat on their pole on the slide so that from a good sample of 2,000 pollen grains it is still impossible to find an equatorial view. Examples are found in sizes up to roughly 50 microns. and include lime, yellow rattle, and hazel.

All pollen grains over about 50 microns are quite spherical. The shape of a great proportion of those below this broad limit bear a close approach to the spherical. Thus they are easy to see under the microscope from all angles. Examples include forsythia, weld, lucerne, and bay.

The tall, prolate grains, when crowded under a cover glass, will all lie on their sides. None will stand up. Polar distances of all grains examined from the umbellifer family are about double the equatorial diameter. Examples, other than the umbellifers are: wild cabbage, marjoram, pulmonaria, comfrey and ribwort melilot and some others in the borage family.

The polyads are clumps of grains, formed in fours and remaining intact en-route to receptive stigmas. Each grain, within a clump, is held next to its neighbours with each of its three corners in touch with one of its neighbour's corners. This is clearly shown in the drawing of Great willowherb. Those of the Ericas, however, appear tightly compact and more like spheres.

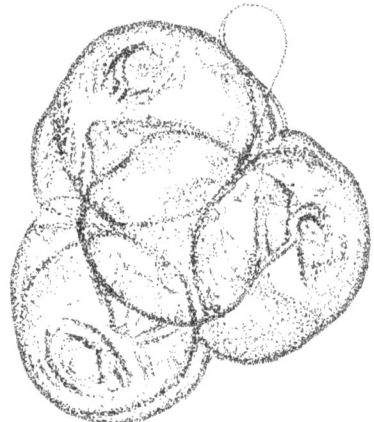 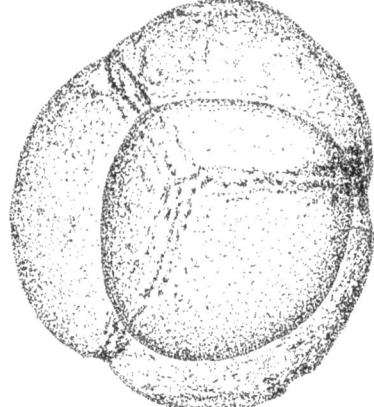

Fig 1. Great willowherb polyad pollen grain (left) and the more compact bell heather grain (right).

Form: the male genetic contents are commonly enclosed within two skins. The inner 'intine' is a thinner, flexible film and the outer 'exine' a harder, less flexible, protective cover. Some pollens have slits (furrows) and some have pores. Others have both with the pores always in the centre of the furrows. Furrows can open a little to allow expansion when the grain absorbs moisture. Most of the grains of size up to 40µm have three furrows spaced equally around the equator, aligned to point towards the poles. The pollen tube, carrying the genetic material to the ovary, emerges from one of the pores of the grain.

Minor differences in the structure of these apertures, seen better when the grains are stained, help with identification of origin of the pollen.

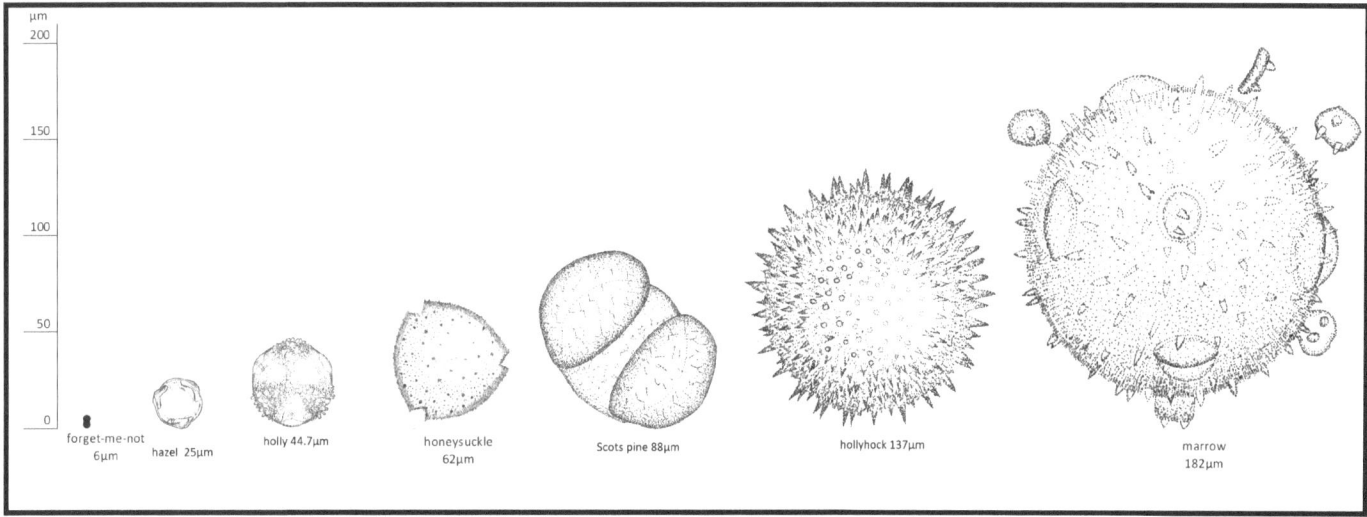

Fig 2. Relative sizes of different pollen grains: forget-me-not is among the smallest; hazel is a reference for size, holly and honeysuckle are larger than average; scots pine turns up unexpectedly; hollyhock and and marrow are among the largest grains.

Collecting pollen

Pollen should be collected from plants on a dry day and placed in a paper bag or envelope. Ensure you have permission to collect if they are not your own plants, and observe the country code. Always record the pollen colour, date and place of collection.

There should be plants in flower, with pollen available, at all times of the year. The easiest grains to collect are on those anthers which are held aloft on stalks (filaments). Examples include: winter aconite, hawthorn, bramble, hollyhock. Other easy pollens to collect are from the catkin bearing plants: hazel, willow, alder, birch, oak, walnut. Also the profuse pollen from yew ~ week 9; and ivy ~ week 39.

Some common pollens with interesting or unusual forms to look at include:

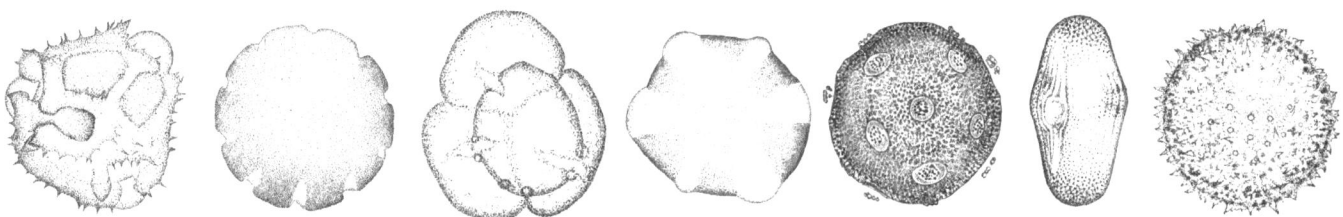

Fig 3. Left to right, not to scale: dandelion, borage, ling heather, rosemary, chickweed, cow parsley, mallow.

Various pollen forms under the microscope

Some pollen grains are quite transparent. Thus under the microscope, by refocusing to the greater depth, the detail of the far surface, looking through the grain, is as clear as that seen on the near surface. The illustrations of a harebell grain show two such views of the same grain, and thus one can easily ascertain that there are three pores. Other pollens from Campanulaceae will have four or a mixture of three or four.

Polyads: A large proportion of plant species dehisce grains bearing three apertures. This is because the grains are formed on the anthers in sets of four[12]. These four usually sit together closely during formation, each one held in contact with the other three, their centres positioned as in the corners of a tetrahedron. There are three points of contact on each, these prompting the formation of three furrows. There is also a strong genetic influence on the number and positioning of furrows and pores. Examples of these tetrad grains may be readily collected from the Ericaceae as all Erica pollens are tetrads. These include rhododendron, azalea and all the heathers. These individual grains never seem to separate. They, and other symmetrical assemblies are known as polyads.

Tetrahedrons: pollen displaying this characteristic comes from a pretty wild flower, yellow corydalis (*Corydalis lutea*) that seeds everywhere, including in mortar crevices in the vertical surfaces of old walls. It has six furrows. The tips come together in threes and actually meet. This happens at four points on the sphere. These points and the furrows are arranged at the points and edges of a regular tetrahedron. The triangular segments of exine are often seen separated.

Fig 4. Yellow Corydalis pollen grain showing tetrahedron formation

Furrows: The most common arrangement is accepted as three furrows, equally spaced around the pollen equator and longitudinally aligned. However, other furrow numbers and configurations are frequently found. Pollens having but one furrow are produced by all the lilies and irises.

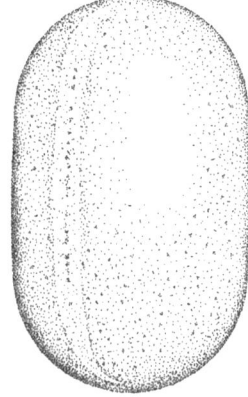

Fig 5. Sainfoin pollen grain showing the longitudinal furrow from pole to pole

Among the well known flowers having six furrows are rosemary (*Rosemarinus officinalis*), balm (*Melissa officilalis*), *Nemesia strumosa* and purple loosestrife (*Lythrum salicaria*). Quite a few have four, or a mixture some with three, some with four, borage has ten.

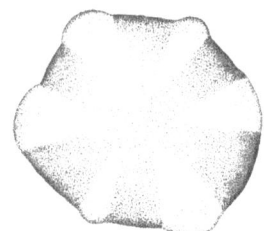

Fig 6. Rosemary pollen grain showing six furrows

Many pollen grains, around a third of local flora, look very much the same, in that they are spherical and have three furrows. This makes them difficult to tell one from another. Subtle differences such as size, surface texture, shape of the furrows and the presence of pores can distinguish about half of these. Identification of the remainder can emerge from clues of pellet colour and the recorded date of harvest. The knowledge of expected flowering dates and the published precise colour charts[2, 3] of the pellets of many plant species can aid pinpointing a pollen source. Some of those stubborn ones still outstanding can at least be slotted into their correct genera or families.

Unique characteristics such as spines or triangular or boat shape outlines are displayed by many. Some from the willowherb family and Ericaceae have long, thin trailing strands. Those of the evening primrose tangle and hold the grains together in clusters.

Most of the bigger grains, hollyhock, mallow and hibiscus etc have around 24 to over 100 spines and pores.

A regular covering of broad based spines is almost certain to herald a member of the Asteraceae, also known as the Compositae (daisy family). Dandelion, golden rod, marigold and michaelmas daisy are all embellished with some variation of these broad based spikes. There are a few exceptions to this including cornflower, which is insect pollinated, and mugwort (*Artemisia vulgaris*) which doesn't have spines, nor even anything that could legitimately be called a flower. Mugwort is wind pollinated.

The illustration of groundsel pollen shows the larger part of a furrow. It will be found wide open or closed depending upon the amount of water absorbed, which can swell or shrink a grain. It exhibits the furrows and spiky protrusions characteristic of the daisy family.

Hairs: some associated hairs may be seen, when using the microscope, among the grains of verbena hybrids. These hairs can appear as a string of beads. They actually comprise a string of closely fitting bulbous swellings, often over fifteen long, and slightly squared. These hairs crowd across the centre of the corolla tube opening. The bee has to push its proboscis through this complex obstacle when reaching for the nectar, and it is believed that the vibration caused by riding these knobbles is enough to release the nearby pollen.

Jelly: the grains of the bay tree (*Laurus nobilis*) show a thick layer of seemingly transparent jelly underlying the exine; the translucent core clearly delineated beneath. The exine itself is decorated with a polka dot array of about 200 minute blobs. These grains exhibit no sign of furrows or pores. Pollen tubes seem to originate from the surface of the core.

Chickweed, a weed of cultivation, covers a large area with its foliage. The tiny white flowers are insignificant. The pollen grains have eighteen pores but no furrows. They also seem to be peppered all over with tiny dots, particles attached to the surface. Those that occur on the pores tend to collect on the pore centres. Those shown on the illustration are borne on the crown of each dome of intine bulging through the pore aperture and hold the cluster of particles aloft, well above the surface level of exine surrounding the pores.

Family characteristics and exceptions

Whilst some family resemblances can give clues to the pollen source, many samples are likely to be from more than one source; and of course there are always exceptions within families. Examples of family characteristics:

Asteraceae or Compositae (The daisy family) Huge family of asters, daisies, sunflowers and many flowering plants.
These, with few exceptions, eg cornflower and mugwort share a characteristic regular covering of broad based spines.

Balsaminaceae (The impatiens family) The oblong shape of the pollen grain is peculiar to the impatiens family.

Betulaceae (The birch family) a large family of birches, alders, hazels, hornbeams, hazel-hornbeam, and hop-hornbeams.
The pollen grains are not identical but have recognisable similarities in outline and features.

Boraginaceae (The borage family) eg alkanet, borage, forget me not, pulmonaria
While varying a lot in size from tiny (6μm) forget-me-not to average (>30μm) borage and pulmonaria, the pollen of this family share the elongated dumbbell shape.

Ericaceae includes the heaths, heathers, epacrids, rhododendrons, azaleas and blueberries.
These plants have tetrad pollen forms and this family thrive in acid soil.

Poaceae or Gramineae (The Grass family) includes cereals, bamboos and wild and cultivated grasses.
Small, spherical pollen for wind dispersal.

Umbelliferaceae or Apiaceae (Umbellifers) (The celery, carrot or parsley family) includes saxifrage, cow parsley and many other aromatic flowering plants.
The pollen is typically an elongated oval shape.

The exine

The tough, protective, outer surface of the pollen grain is often elaborately sculptured in distinctive patterns[6, 7, 8]. These can be seen in great detail in scans using electron microscopy. Under the light microscope a smooth surface, netting patterns, and surface spines are some different characteristics which can still be observed albeit at this lower magnification.

Difficult identifications

Tiny, and with deceptively elusive flowers is pollen from buckthorn (*Rhamnus cathartica*). It is visually indistinguishable from that of yew (*Taxus baccarta*), so much so that they were confused in a national honey survey.

It is technically possible to identify some of the difficult pollens by scrupulous observation and recording. Suggested criteria would be:

1. Polar field. This the distance between the tips of the furrows approaching one pole, expressed as a proportion of the equatorial diameter
2. The tendency to an oblate or prolate form.
3. Pore diameter
4. Size and characteristics of surface granulation elements or net cells
5. Transparency

Individual identification, if this method can be used at all, is just not feasible visually for pollens extracted from honey. The only way when all the many grains are intermixed, as they are in examples centrifuged from honey, is to learn to tell all the likely ones, and if the information is available, compare with known flora in the area from when and where the honey originates.

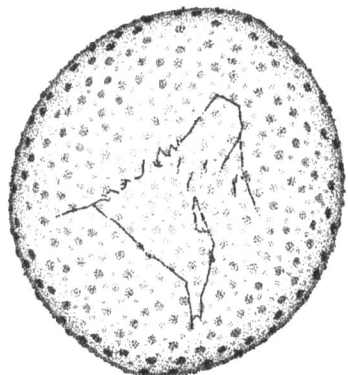 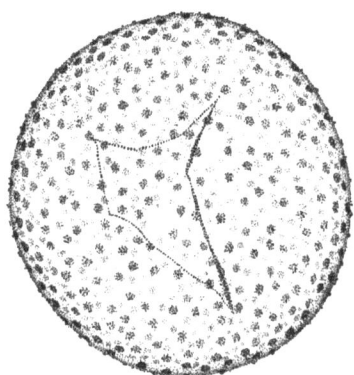

Fig 7. Buckthorn pollen 25.2µm (left) is difficult to distinguish from yew pollen 25.8µm (right).
The image of the core changes with each increment of the fine focus.

Collecting pollen samples direct from the flowers

Endless physical arrangements of florets in cluster, spike or catkin form are encountered. Some anthers are held high and clear making them easy to access. Others are secreted deep inside corolla tubes or other covert corners such that it is a wonder how the bees are able to access them at all, snapdragons being a classic example. Some challenges are found with the pollen of forget-me-not; clover; gorse, and other legumes.

Collecting by 'touch on' to the microslide. A great number of common flowers hold their pollen well out in front. Examples include blackthorn, fuchsia, hibiscus, honeysuckle, hypericum, passiflora and spiraea. Touching the anthers with a microslide, prepared with the merest smear of glycerine jelly secures the sample, ready for staining and mounting. This method has the advantage of avoiding overcrowding, where too much pollen forms two or more pollen thicknesses under the cover glass.

It helps to harvest just one, two, or a bunch, of anthers in order to take as little clutter of the rest of the flower as possible, other than the pollen itself. Some are so easy, e.g. lilies, honeysuckle, hypericum and willow in which the anthers are held clear and prominent. Lilac and buddleia flowers secrete their anthers halfway down the nectar tubes. The most difficult are often amongst the smallest. These include buckthorn, ceanothus, gorse, green alkanet, chickweed, goosegrass and some of the labiates.

More difficult pollen. Many plants have tiny flowers and need laborious pulling apart and separating of enough diminutive stamens to gather sufficient pollen to give good representation when mounted. Each flower species has pollen tucked inside in its own way making it difficult to liberate the anthers within. Rarely the work may be further hampered if the flower is small and loaded with nectar. In these wee sizes it is like glue, and an additional task is removing plant fragments, whether needed or to be discarded, from the tips of the forceps.

With experience comes the development of techniques for collecting and separation. The shaking method, i.e. using a small glass bottle of about 8ml, is a quick way of collecting nearly all pollens. Some plants like mahonia have many small flowerlets on one spike. Grape hyacinth and redshank are almost impossible to dissect as the individual flowers are so tiny. A collection of these discrete flowerlets, or bundles of stamens cut from flowers like hypericum, and catkins can be dropped into a bottle of neat propyl alcohol (propanol). A shake is hardly needed. The pollen is dislodged and becomes suspended in the liquid. It can be stored like this and mounted later. In this way enough grains can be amassed to fully furnish two or three microslides with 1000 grains each.

A useful way to collect from spearmint or redshank, which have tight spikes of small flowers, is to cut some spikes with stalks and to stand them in a pot of water. Some flowerlets protrude and open. These can then be pulled off as they protrude and be added to the alcohol pot. Thus one can accumulate a reasonable sample over about four or five days. This approach doesn't work with rosemary but it pays to try and see.

Useful kit for pollen collecting:

1. Two to five 8ml glass bottles with polythene clip-on, or screw-on stoppers.
2. A 15 or 20cc glass, screw top bottle containing propanol
3. Five prepared microslides
4. A pair of fine scissors
5. A small, sharp knife
6. Small stick-on labels
7. Note book or paper
8. Pencil, ball-point pen
9. Watch maker's eyeglass (Fig 8), or hand lens
10. Small cutting pliers
11. Paper bags
12. Plastic shopping type bags, very useful also to kneel on
13. Wild flower book for identification
14. A prepared plan in what to expect to find and to collect.

Avoid carrying the bottles in the same bag as items that might be despoiled by the alcohol eg book, cameras or food.

Fig 8. *Watch makers eyeglasses x2.5, x5 and x10 left to right.*
These are held by scrunching the skin around one eye usefully keeping both hands free.

Microscopy
Microscope power
A compound microscope of 100 times overall magnification is needed to see pollen grains.

With a x100 microscope you will be able to just see the grains of pollen. Using x400 you will be able to see some identifying features. You will need x1000 to see the level shown in this book.

All the drawings illustrated in this book were made with the use of a dated, university redundant, Carl Zeiss Standard RA Microscope.

Stereo Microscopes
A stereo microscope greatly improved the quality of my mounts. It proved useful for examining prospective cover glasses for cracks and edge defects. It came into its own because of the working space under the objective lenses in enabling microscopical rubbish to be removed using a mounted needle. The needle point is touched on to each spurious item which sticks to the needle point by electrostatic attraction, and is thus readily removed.

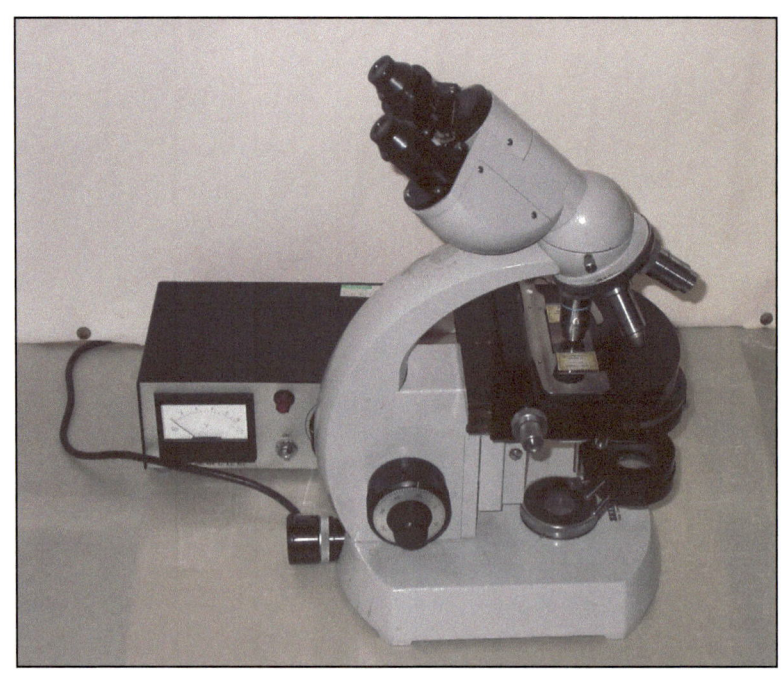

Fig 9. Carl Zeiss Microscope dated approximately 1930. A x15 pair of eyepieces, borrowed from a Russian microscope, is often used to give a magnification of 1500.

Fig 10. A DIY clean up guide shows if the grains are within in the area where the cover glass will go. Sized 115 x 58 x 5mm high, the rectangular slot holds the slide. The cellophane film carrying drawn centreing lines, is stuck underneath. Has space above to enable hairs and micro-rubbish to be removed

Coupling the camera to the microscope

The camera I started with was a Nikon Coolpix 995, popular among microscopists because it comes with a thread around the lens. This enables the camera to be fixed, with the help of a coupler, onto the microscope. Now well out of production. I later bought, and use, a Nikon Coolpix 4500. This design, a lookalike based on the 995, retains many of the original features.

Setting up a compound microscope

A suitable mounted sample is first set up in the microscope so that it can be viewed. Consult relevant literature or on-line guidance. To summarise:
1. The 'lamp' (or 'field') iris is opened so that it does not restrict the field viewed and the specimen is focussed.
2. The field iris is stopped down and focussed on the specimen; then opened up again to get a full field of view.
3. Alternatively, a pencil point can be held just in front of the light source where the field diaphram, if present, is situated. It is waggled so that the image is seen through the eyepiece to move with it. The condenser is then adjusted for height to bring the image of the pencil point into focus, superimposed upon that of the sample. The pencil point technique is good if your microscope doesn't have a lamp- or field- iris.
4. An eyepiece is removed and the back of the objective lens viewed. The sub-stage iris (or 'aperture') diaphragm is then adjusted to reduce the illuminated disc to 85% of its full diameter. This controls the size of the cone of light being supplied from the specimen to the objective lens. It changes the contrast observed. Manufacturers sometimes supply a microscope telescope to temporarily replace the eyepiece, making this view easier.

As different objective lenses are swung into use this setting up may need readjustment.

The Condenser

Setting up the condenser can be more help than just using it as a mirror.

By cutting a strip of tracing paper. It should be one cm wide by ten cm long. By slipping it over the top of a mounted microslide it will show the pattern of the beam lighting the sample.

More importantly, it will reveal how the shape changes with slight movement of the condenser height and will enable observation of a corresponding change of beam shape.

The conical beam shape can be adjusted when a central round dot is set up at the sub stage iris level. The resulting hollow pyramid of light gives an image that is perfectly ringed.

Fig 11. Camera, set on the microscope showing a coupler between the two. It was used to take many pollen grain photos.

Preparing the pollen grain for microscopy

Preparing Slides
Preparation of slides for the purpose of appreciating the different morphologies, sizes etc of pollen grains takes time and practice. The ideal is to prepare a slide, under a cover glass of 22mm square, with 1000 to 2000 pollen grains, a sufficient number to be able to find the grains at the desired magnification.

Getting the pollen on to the slide
Pollen is found as clumps on the anthers of flowers. These anthers are cut and dropped into a small 8ml bottle. About 1ml of propyl alcohol (propanol) is added and the bottle is shaken. This action releases the pollen grains which remain in suspension. The grains are heavier than the molecular weight of propanol.

Using a fine syringe take a small quantity of pollen from the bottom of the bottle and swirl it gently in a watch glass. This circular action concentrates the pollen into the centre, enabling the relatively clear liquid near the edges to be syringed away and returned to the bottle. Further pollen can be added if required. The quantity to achieve the ideal spread on the slide will become obvious with experience.

The cluster of pollen at the centre of the watch glass, with little of its original liquid can then be drawn off and placed in the centre of the a microslide and spread with a needle point to the size and shape of the intended cover glass.

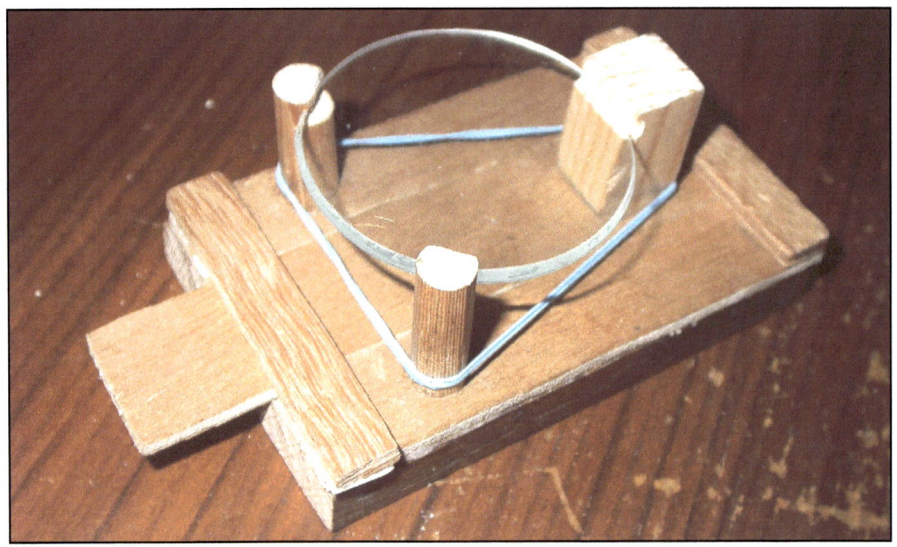

Fig 12. A DIY watch glass holder. Grips a 2 inch glass.

The pollen on the microslide is left for two or three minutes to dry. Then the microscope can be used to assess the ideal quantity and spread. A paper mounting template (Fig 14) shows if the grains are within the cover glass area.

The grains on the slide can be stained, and then stain left on the background can be rinsed away using propanol. Despite the propensity of pollen to stick to the face of the glass slide during processing, rinsing with water must be avoided as this washes it all off.

Stains are available in several colours and enhance the visibility of detail. They are used to counteract serious reductions of contrast at the high magnifications used. Choice of colour suits personal taste but once selected it is recommended to remain using the one colour.

Hydration, exposure to damp air for five minutes immediately before mounting, enables the grains to absorb moisture making the exine stretch. This gives the grain better form. Without a hydration process too many samples appear deflated. Unfortunately not all species are found to respond well to this.

The pollens of some species eg evening primrose, tend to clump together to the extent that individual grains can't be seen under the microscope. Stirring of the sample, just as the last of the propanol is evaporating, can help to break up the clumps.

Fig 13. Mounting press. Holds cover glass down securely. Hinged on the right and held down with weights or two elastic bands. Pressure pad is of silicon rubber, 2 x 2 x 1.5 cm deep. Surplus mountant will not stick to it.

Mountants

Mountants have a refractive index closely similar to that of glass (1.53). That of air is 1.0. They are used to replace the air under the cover glass, encase the grains and keep them firmly in place and minimize reflections from glass surfaces.

Two mountants are used widely in microscopy. Each has advantages which the other lacks.

Glycerine jelly is recommended for beginners, with the staining and mounting done together for those applications where results are needed quickly. This mountant sets within minutes of cooling and mounting so the exercise can be completed within a course session.

Glycerine jelly mountant comprises gelatin, glycerine (glycerol) and water, usually in the ratio 7:50:100 respectively. It becomes a stiff jelly melting at about 40°C, reverting to jelly again at room temperature. A commercially available sealant around the cover glass edge is recommended to keep it in place as otherwise it leaks over time.

Canada balsam is a more permanent mountant, usually supplied diluted 50% in xylene (xylol). A few drops of xylene added to the balsam bottle after using it for mounting helps replace usage and evaporation. Canada balsam can be applied at room temperature. However, it relies on evaporation of the xylene to set which can take two to six months as the xylene is trapped in the confined space under the cover glass. This can be speeded up to about a month by using a warming plate and cleaning up surplus oozings with toluene but this is risky and can disturb the sample, introducing damage and early leaking.

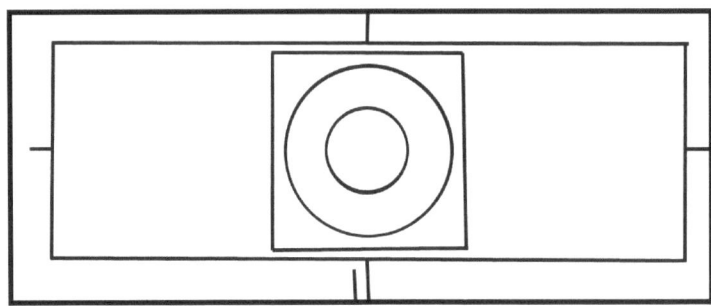

Fig 14. Paper mounting template cut to 3.5 x 8.5cm.

Making the drawings

Collecting pollen grains, as they become available during the year, provides the starting material. The quality of the mount, and the size and form of the grains influences the selection of species to view and draw. The unusual and interesting grains are fascinating to discover, and it is gratifying to develop the experience in recognising grains from families well known or of economic importance. It also becomes obvious that the entire floral spectrum could not be covered in a book like this.

In preparation to drawing any grain, it's a good idea to view many of the grains of that species on the microslide in order to take in a good assimilation of the form for that species and select a typical specimen to draw. Having selected a suitable grain, I photograph it using a camera with a microscope coupler.   I make prints selected from several levels of fine focus to trace the beginning of the final artwork. This gives me an outline plan, and aids memorising the view back and forth from the eyepiece to making the drawing. From the photos, tracings, and direct viewing, I fill in the detail employing shading techniques to add the impression of shape to the final drawing.

The illustrations can show much that the microscope cannot. For example, the drawings are entirely in focus.

My drawings usually treat the grain as a solid opaque object showing mainly the surface detail rather than showing the grain as a translucent body with contents as revealed by the microscope.

There are software packages available to facilitate 'image stacking'. This is a technique to view a 3D object focusing at successive focus levels, to build up a 3D image. Multi-layer stacking will process a selection of neighbouring layers that have been photographed, pick out the in-focus areas and join them into an all-in-focus whole. Multi-layer stacking does not introduce any more detail than can be derived with ordinary photography. Stacking has an established place in microscopy. It might help one to make drawings, but was not used to make any drawings of the grains in this book.

Bees and Pollen

As well as in attracting any falling wind blown pollen to a nearby stigma, electrostatic attraction also plays a part in insect pollination. Many vectors fulfill this task eg flies, beetles, moths, wasps and humming birds. A few plants trap insects until pollen is transferred. However, the majority of pollination, worldwide, is performed by various species of bee.

As well as being drawn by its colour, including features only visible to humans when illuminated by ultraviolet light, bees can sense the tiny electrostatic field around a flower. Flowers tend to have a negative charge, accumulating a positive charge in the air immediately around them. The branched hairs of the bee have a concentration of lines of electrostatic force around their tips. As the bee flies through the air, friction from charged dust particles removes surrounding electrons, leaving a surrounding tiny positive charge. As the bee approaches the negatively charged flower, pollen flies to the bee so that more sticks to the bee than just by brushing past the anthers. Pollen grains, eg those of nearly all the Compositae, which have a surface structure that bears spines, or points, support a stronger electrostatic force at their pointed tips adding to the natural pull and taking full advantage of the effect.

The reduced negative charge left after a bee's visit renders the flower less attractive to the next bee until the charge, and presumably nectar supply, build up again.

When approaching the stigma in the next flower, the stronger positive charge on the stigma draws pollen grains towards it, thereby delivering the pollen to the receptor so that fertilisation can take place.

However, only a tiny quantity of pollen is needed for pollination. The greater proportion of pollen collected by bees is used to feed their young. Pollen grains are rich in protein as well as containing smaller amounts of carbohydrates, fats, vitamins, minerals etc. They provide essential food for bees, particularly in the early spring when large numbers of workers need to be raised to take advantage of the summer yields.

Fig 15. Hollyhocks produce copious quantities of large pollen grains which liberally coat visiting bees.

Pollen from a trap on a honey bee hive

A pollen trap is a valuable facility for a beekeeper with an interest in pollen. Its main use to the beekeeper is to collect pollen for storing and feeding back to the colony in the late winter months providing food for the spring build up. In addition, it can provide more reliable, information from collected pollen extra to that from pollen centrifuged from honey. Colour can be seen in the collected pellets and the date of the samples can be recorded. In addition the bees habit of constancy, visiting just one species at a time, obligingly makes each pellet up with the pollens of one species of plant. This, unmatched elsewhere in natural history, is a veritable gift for studying the pollen.

Fig 17. Pollen trap on a hive.

That all grains in a pellet, apart from strays, are from one species of flower enables comparison of form between pellets and gives an effective review of the local flora visited. Spaced at say, weekly intervals, a hive set upon a pollen trap on a sunny day for less than 12 hours will harvest an abundance of pellets. Samples thus obtained can easily be dried and kept in small containers, dated at intervals round the year giving an accurate and unique diary of the flowers that the bees have visited, mostly within about a mile radius.

Three such collections of pellets at about three week long intervals in June/July, were sorted easily under a lamp into their colour bands. The range of bright yellows, greyed yellows, pale yellow-browns, various shades of green, orange, reds, deep violet and even black pellets was fascinating, and completely dispelled any prior notion that all pollen was yellow. Some colour bands were dominant on some dates and entirely missing on others. It was quite amazing to observe the drastic proportional changes of these colour bands across only a few weeks.

Fig 16. Pellets sorted by colour, collected (top to bottom) 29th May, 9th July, and 31st July 1998.

Summarised here are the benefits of processing and analysis of pollen derived from a trap:
1. It is readily accessed by beekeepers.
2. The colour and date are there for the recording. This is not so for pollen extracted from honey.
3. An enormous supporting collection of dried pellets, from spot dates at intervals round the year can be readily amassed, already sorted into discrete species. This takes up little space, and a study of these pellets offers much information on the flora within flying distance from the hive.
4. A collector can accumulate knowledge and experience very quickly by starting with trapped pollen. A microslide can easily hold over a thousand grains, These all from one flower species. In spite of a few misshapen ones, which are sometimes included, the dominance of numbers and the different angles of view leaves no doubt with the observer of the general form of that grain. The stray grains that are found in minute proportions in most slides from pellets can also help with identification skills.
5. The numbers of slides that need to be mounted for the microscope enhances the mounting skills. The more you do the better you get at it, for both quality and effectiveness, and mounting from pellets is more easily accomplished than preparation of pollens from flowers.

A pollen trap needs to be fitted and emptied each day of use to avoid the build up of tiny insects, spiders and avoid harbouring wax moth.

The pellets can be isolated by shaking in a kitchen sieve. This removes the smaller, dust size particles. Further unwanted elements can also be removed at this stage.

Fig 18. Drying pot for pellets. Pollen pot inside the outer pot containing silica gel crystals.

The pellets' moisture content means they do not keep well unless dried. Drying modest quantities can be carried out using dry, blue, silica gel crystals to absorb the moisture. Pollen pellets can be placed in a small transparent jar, covering its open end with fabric or kitchen paper secured with elastic bands. The smaller jar is then placed face down in a larger transparent jar containing some of the crystals, and the larger jar then sealed to make it air tight (Fig 18).

Transparent containers enable the colour of the crystals to be checked every few days without disturbing the contents. Moisture removal can be verified by weighing the pellets before and after.
The pellets usually feel more crusty when dried.

By this means one 47g batch of pollen lost 4.8g weight in 4 hours and a further 2.5g in the next 48 hours. Another 10 days took it down another 1.2g after which no further fall was measured. That revealed at least 8.5g or 18.2% of original water content. By experimenting with storage at room temperature versus refrigeration, the silica gel in the refrigerated pot stayed blue throughout, while the crystals in the pot stored at room temperature needed redrying every month or so. Neither pollen sample showed any sign of going mouldy.

Pellets, when fresh, are soft and can be squashed with the fingers. When dried however, they can be quite hard. Pressed between hard surfaces eg between teaspoons, such a pellet will crack and bits will split off. A pellet dropped into about 8 drops of water will break up. Some of the cohesive constituents dissolve and all the grains, around 50,000 are dispersed into the water. This can be done in a watch glass. Occasional cases are encountered in which the grains are locked together. One such case is that of the willowherb family. Long thin strands are attached to each grain. These tangle and tenaciously hold the grains together in clusters.

Stray pollen within a pellet
A visual scan of a whole microslide of trapped pollen, a field width at a time, will frequently reveal a few unexpected stray grains, among a host quantity of 1000 or 2000. Grains of pine (*Pinus sylvestris*) are to be found in many May or June collections and are subject to surprising appearance at other times. Small grains like a close patch of twenty green alkanet (*Pentaglottis sempervirens*) at 13μm were found amongst a host group of 1,500 grains of plum (*Prunus* sp.) in April. Many other stray grains have been spotted. Harebell (*Campanula rotundifolia*), yew (*Taxus baccata*), daffodil (*Narcissus* sp.) *Mahonia lomariifolia*, and even Great bindweed at 100μm all in a pellet of Compositae in July. It is necessary to search through a pollen slide slowly strip by strip in order to find, identify, and if necessary ignore these odd ones.

These strays arrive adhering to the backs of insects other than honey bees. Bumble and solitary bees, hover flies, blow flies, beetles and many more visit the same flowers as the honey bees, yet unlike the honey bee, they criss cross between flowers throughout the day haphazardly depositing whatever stray pollen they happen to carry.

Rarely a trapped pellet will show a mixture of two quite different grains. The reason could be that as plants eg great bindweed (*Calistegia sepium*) dehisce their pollen only in the morning and others eg evening primrose (*Oenothera erythrosepala*) open in the evening, the bee transfers its allegiance elsewhere for the remainder of the day.

Although pollinated mainly by moths, nevertheless some of evening primrose grains seem to get into the hive. The composition of these pellets may reveal such clues. As an example, one of my trapped pellets prepared for the microscope was quite black. When viewed it clearly revealed two quite different grains, therefore from two species. The larger one I thought at first was from daffodil, but later realized it was too small for that and found the size, at 28μm, was more usual for snowdrop. But that was not conclusive as there are many cultivated varieties of daffodil, and the pellet was trapped in late March, unusually late for snowdrops.

This is where a reference library from a range of samples within a species, in this case the Amaryllidaceae, to which both snowdrops and daffodils belong, would prove useful to aid identification.

Fig 20. Centrifuge. Can be used to spin down and separate pollen from honey.

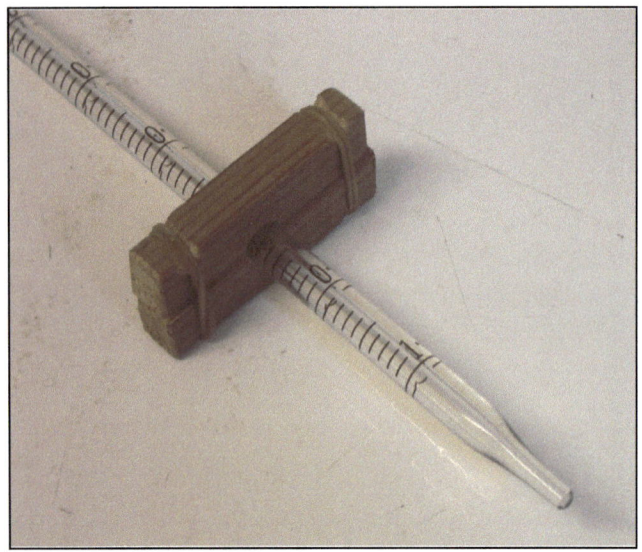

Fig 19. Slide bar. Two identical wooden pieces, shaped to clasp onto the graduated pipette. Held in place with two elastic bands. This makes a useful depth gauge.

Fig 21. A DIY centrifuge bucket stand holds centrifuge buckets. It could be made from thick cardboard.

Extracting pollen from honey for microscopy

Pollen analysis can also be used to help identify the source of honey and wax.

Illustrated, are a wooden slide bar which, slipped over a graduated glass pipette, (Fig 19) converts it into a depth gauge, and a DIY stand (Fig 16) to hold four buckets upright. Used together these are without match for convenience. The graduated pipette or a glass rod, held by the wooden slide bar, used as a depth gauge in the bucket , so that the narrow end touches the bottom with the slide bar butting against the bucket top. The bar is then eased downward by about 5mm so that all but 5mm of water can be carefully sucked or siphoned away.

Procedure

Take 55g honey and mix with 165g water to give 220g of dilute honey.
Place 55g into each of 4 centrifuge buckets.
Spin at 5000rpm for 20 minutes.

Take off ~50g from each bucket, leaving 5g x 4 = ~20g.
Shake each bucket and empty into one cup, add water to make up to 220g.
Stir and place into 4 buckets. i.e. 4 x 55g.
Spin at 5000 rpm 15 minutes.

Take off ~50g from each bucket, leaving ~5g x 4 = ~20g.
Shake each bucket and empty into one cup, add water to make up to 220g.
Stir and place into 4 buckets. i.e. 4 x 55g.
Spin again at 5000 rpm for 15 minutes.

Take off ~50g from each bucket, leaving ~5g x 4 = ~20g.
Shake each bucket and empty into one cup, add water to make up to 55g.
Stir and pour the 55g into one bucket.
Mark it as the sample, then add water into a second bucket, to equalize and balance the centrifuge.
Spin again at 5000 rpm for 15 minutes.

Remove and discard the top 50g from the sample bucket.
Shake the remainder and pour into an 8ml glass bottle.

Successively transfer 3ml into a watch glass. Swirl and return or discard the surplus liquid to concentrate the pollen.
Deposit the pollen on to a glass slide.

As described previously (p14 - 16), stain, moisturise, and mount in Canada balsam.

Pollen in Honey

Much honey for sale is a blended mixture of honey from more than one source, and/or more than one season. Not all pollen is distinctive so it's not always easy to accurately identify the floral source of the honey from microscopic examination of pollen grains from honey samples.

Honey bees have the habit of 'constancy', visiting only one flower species until that source runs out.

So called monofloral honey, collected from hives close to, taken to, or whose bees are assumed to be visiting specified crops *eg* oil seed rape, borage, heather etc would be expected to contain pollen grains from that particular plant.

Some hay fever sufferers seek out honey produced from allergen flowering time in the area they live in, hoping the ingestion of the pollen will aid in desensitising them and reduce their seasonal suffering.

There are commercial companies which specialise in various honey analyses including pollen content.

Pollen Drawings

Winter Aconite

Eranthis hyemalis 30.0μm
Ranunculaceae

Attractive, perennial ground cover plants in the buttercup family. Five or so carpels sit at the centre and form a star shape as the seeds develop, several to each arm. They disperse around the area of the plant creating a rapidly spreading carpet. Flowering in winter, they are a good source of early pollen, so a useful addition to the wildlife garden.
The pollen grain has three furrows.

Agapanthus

Perennial summer flowering bulb, almost hardy in the UK. The flat, boat shaped pollen grain is characteristic of the Amaryllidaceae.

Agapanthus africanus 48.0µm
Amaryllidaceae

Ageratum (Floss flower)

Ageratum houstonianum 25.8μm
Asteraceae (Compositae)

The typical spiny daisy form pollen grain with three or four pores.

Alder

Alder is monoecious, ie both male and female flowers are found on the same tree. Male catkins are yellow and pendulous, measuring 2-6cm. Female catkins are oval and green, three to eight on each stalk. Mainly pollinated by wind, the small, cone like female catkins stay on the tree all year round. They open up to release seeds, which are then dispersed by wind and water.
Alder, like hazel, belongs to the birch family Betulaceae, and similarities can be seen in the shape of their pollen grains.

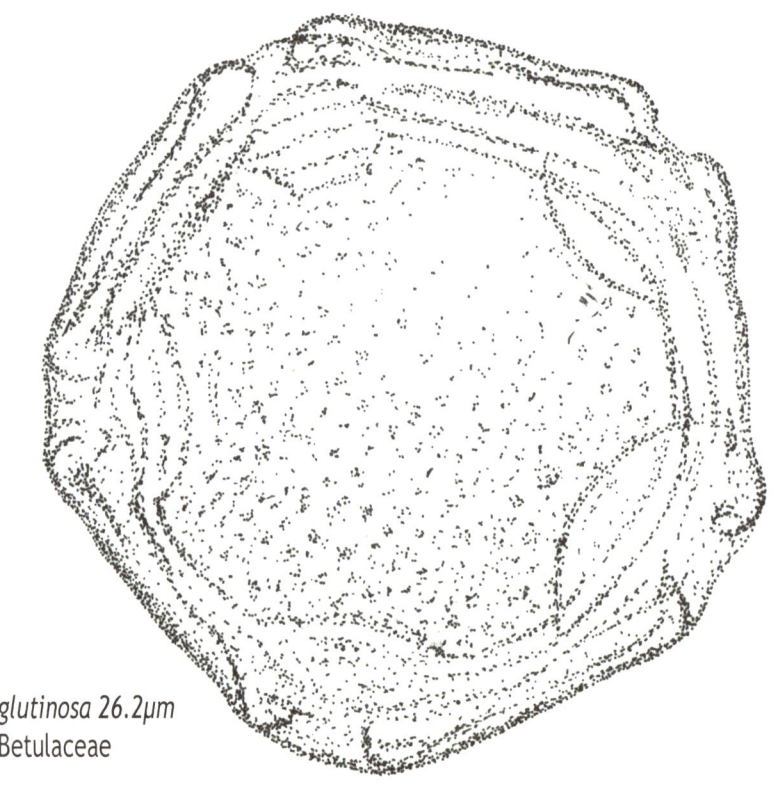

Alnus glutinosa 26.2µm
Betulaceae

Green alkanet

Slightly elongated pollen grain shape with furrow indentations faintly visible.
The plant grows abundantly in the wild.
Although a member of the borage family there is no visble resemblance in the pollen grains.

Pentaglottis sempervirens 12.0μm
Boraginaceae

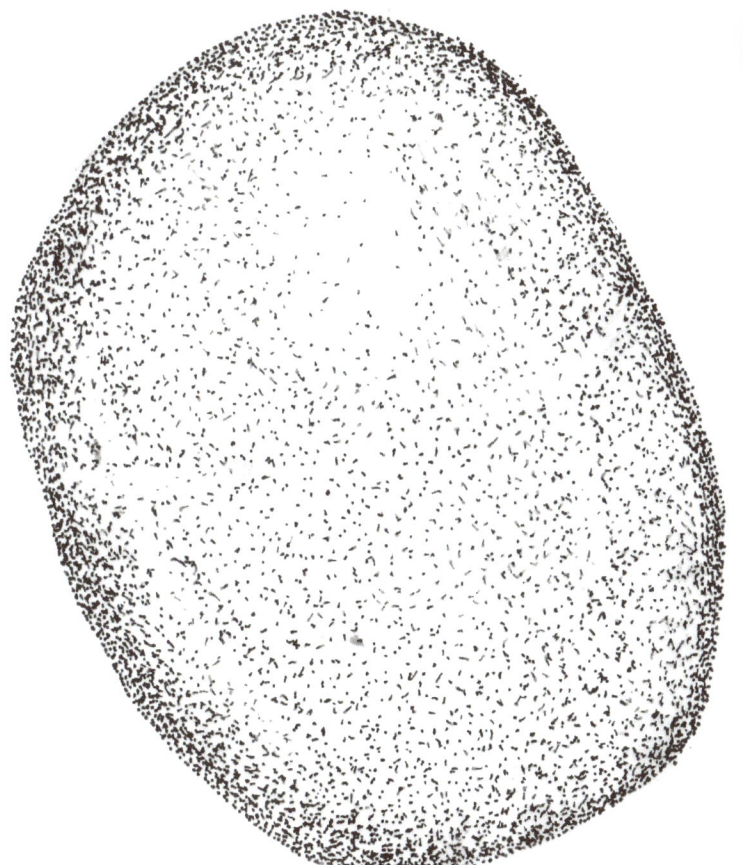

Sweet alyssum

Low growing spring/summer flowering plant often used as ground cover. The polar view of the pollen shows the three furrows and surface texture of the exine.

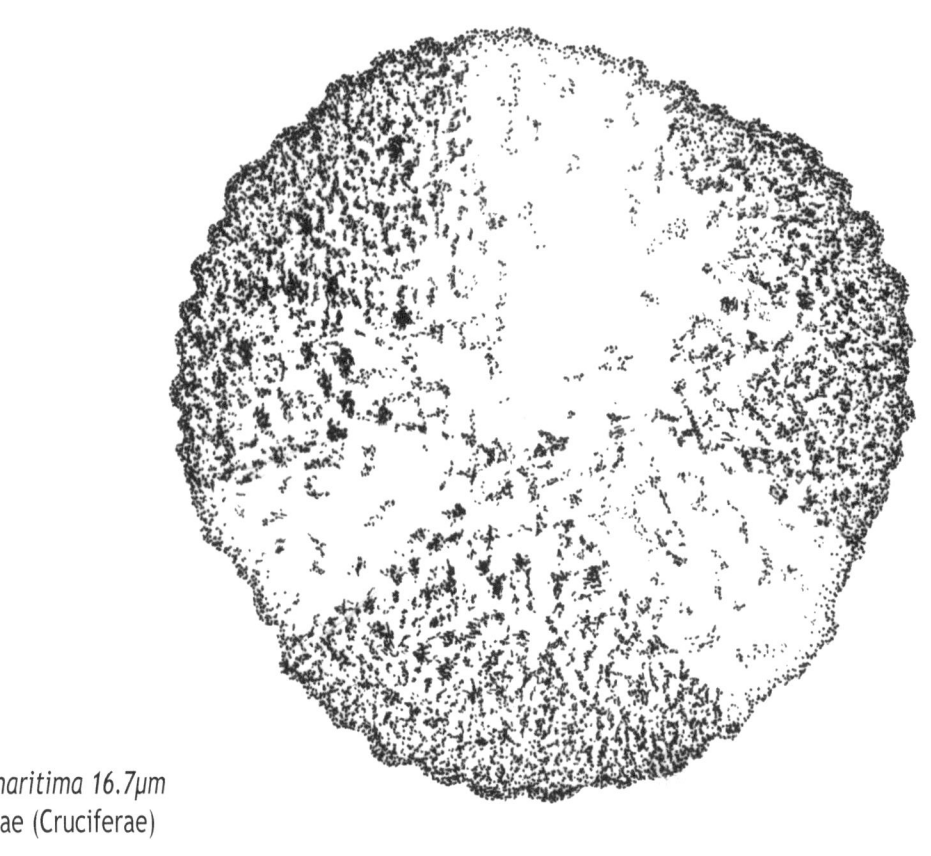

Lobularia maritima 16.7μm
Brassicaceae (Cruciferae)

Amaryllis

The single furrow gives the pollen grain a flat sided, boat like shape. The single furrow is typical of the Amaryllidaceae.

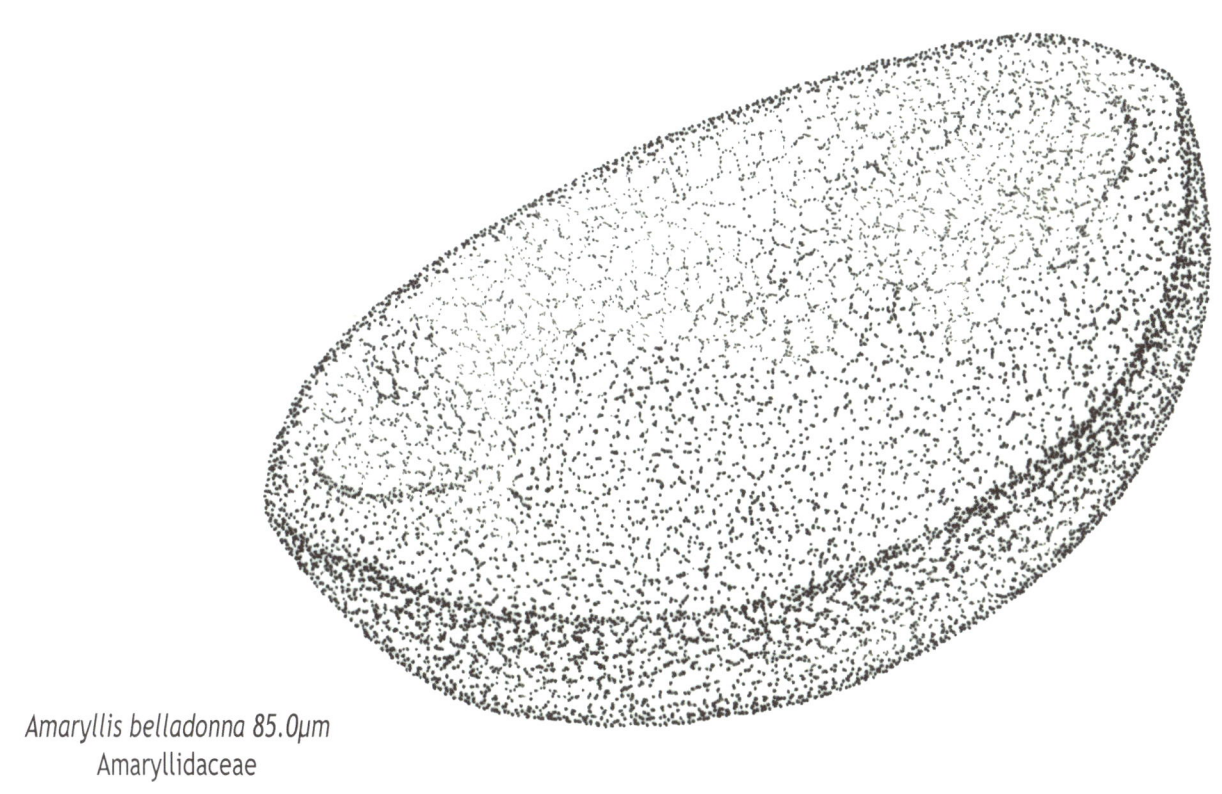

Amaryllis belladonna 85.0μm
Amaryllidaceae

Apple

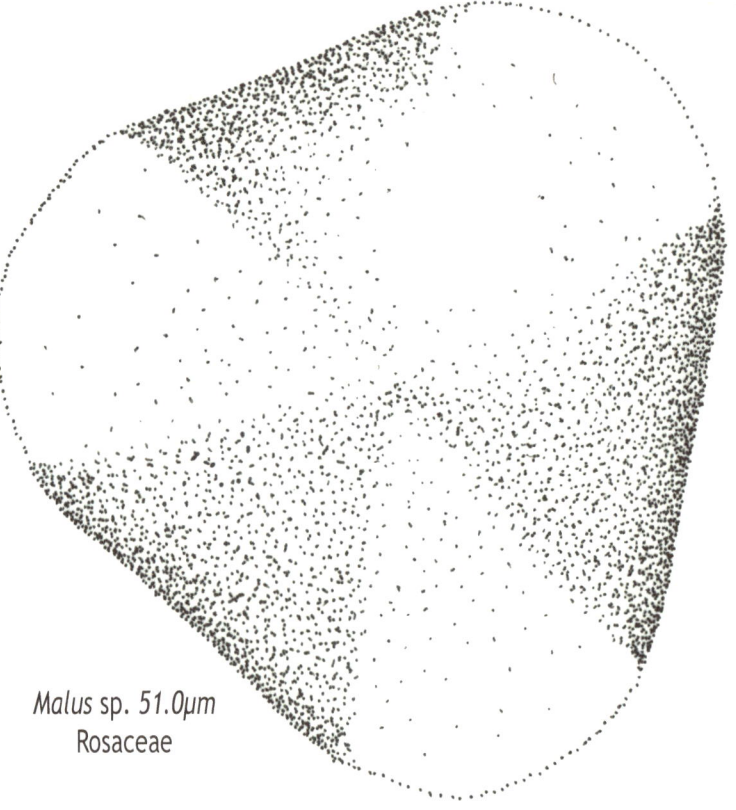

Malus sp. 51.0μm
Rosaceae

Although the flowers of apple trees have both male and female components, almost all have the anthers maturing before the ovaries, making them self sterile, so needing cross pollination. They are visited by insects for both nectar and pollen.

Aquilegia (Granny's bonnet)

Polar view of the pollen grain, which is elongated pole to pole. The furrows do not meet at the poles, giving the grain a flat topped appearance.

Aquilegia vulgaris 22.0μm
Ranunculaceae

Arabis

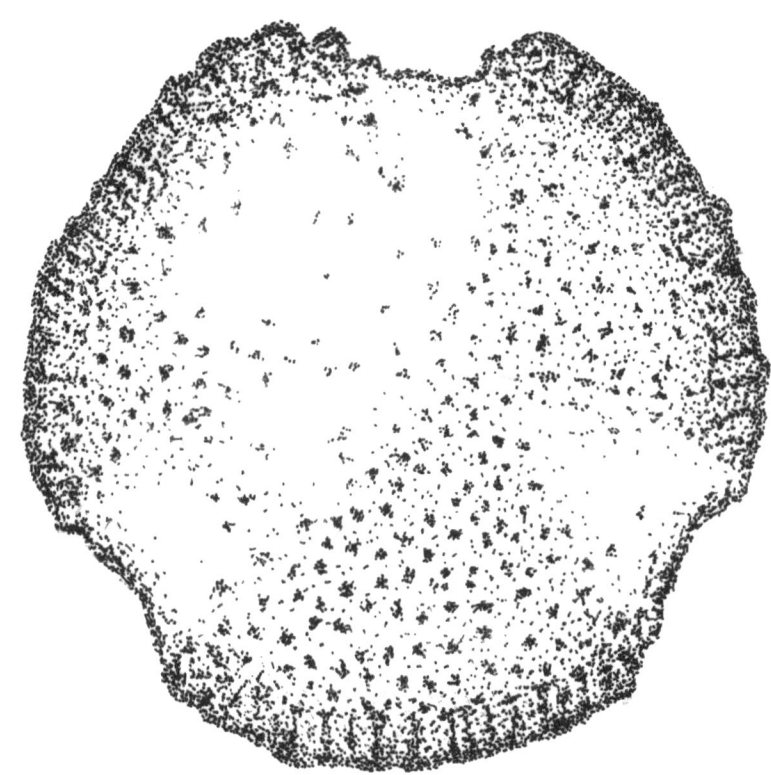

Arabis caucasica (Arabis alpina, Arabis albida) 20.0µm
Brassicaceae (Cruciferae)

A famliar garden 'edging ' plant known as rock cress.
Polar view of pollen which tends to be elongated, having three furrows giving indentations along the length.

Artichoke (globe)

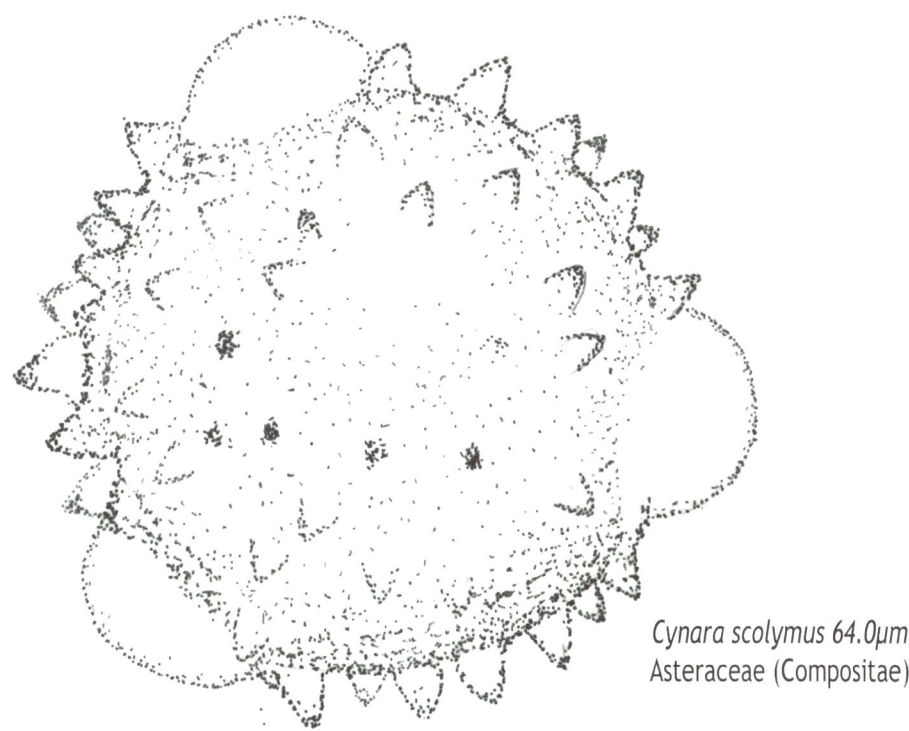

Cynara scolymus 64.0μm
Asteraceae (Compositae)

The large artichoke head consists of many individual purple florets. The spiny surface of the pollen grain helps to concentrate the electrostatic field which attracts the grain to foraging bees. The intine can be seen bulging through the three pores.

Arum

Most Arum lilies do
not reward their pollinators with nectar
or pollen.
Insects are enticed into performing
pollination by the flower emitting an
irresistable chemical scent attracting
insects such as flies and beetles. The
flowers entrap and imprison the insects
which pollinate the female flowers as
they escape. See also lords-and-ladies,
the familiar wild arum with similar
flower construction.

Zantedeschia aethiopica 38.3μm
Araceae

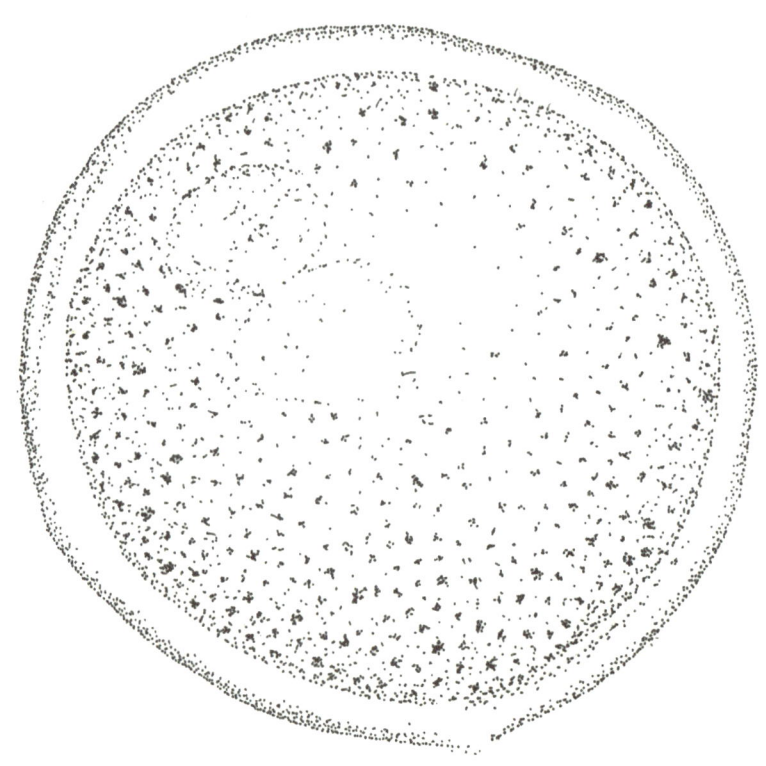

Ash

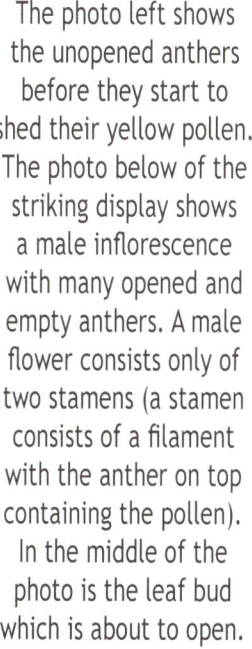

The photo left shows the unopened anthers before they start to shed their yellow pollen. The photo below of the striking display shows a male inflorescence with many opened and empty anthers. A male flower consists only of two stamens (a stamen consists of a filament with the anther on top containing the pollen). In the middle of the photo is the leaf bud which is about to open.

Fraxinus excelsior 27.3μm
Oleaceae

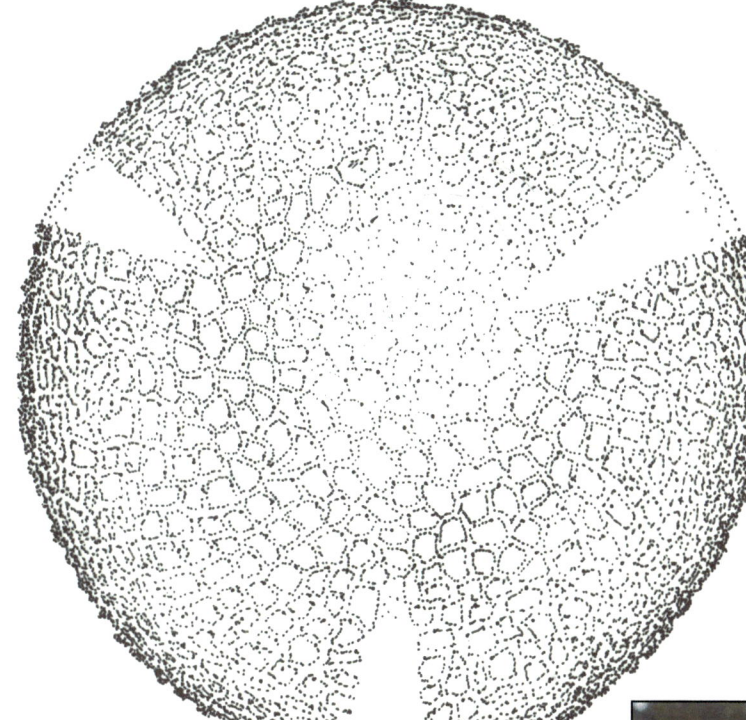

Photo © Sally Dunn

Ash has a complete spectrum of sexuality between predominantly male or female trees, often with a mix of hermaphrodite flowers on the same or different branches or even in the same inflorescence as the main sex.
This polar view of the pollen grain shows the pattern of surface netting. The furrows are sharply defined and short, with no germ pores.

Balm

The pollen grain has six furrows in common with many of the Lamiaceae.

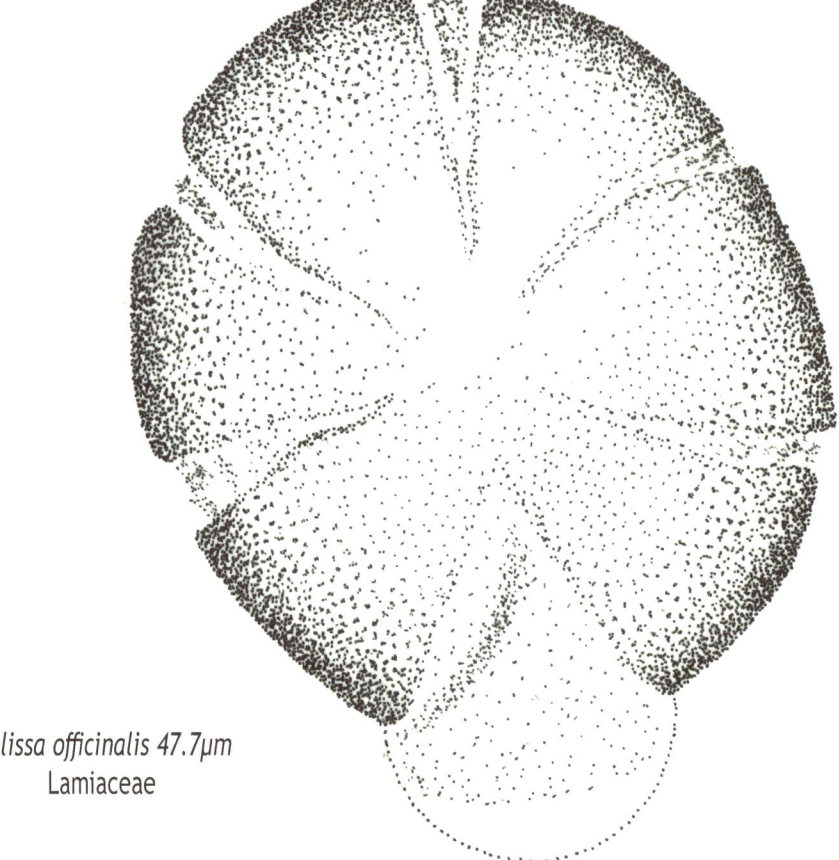

Melissa officinalis 47.7μm
Lamiaceae

Himalayan Balsam

An invasive plant of the impatiens family with its characteristic elongated shape and showing a furrow at each corner.

Impatiens grandulifera 32.1μm
Balsaminaceae

Photos © Sally Dunn

Small balsam

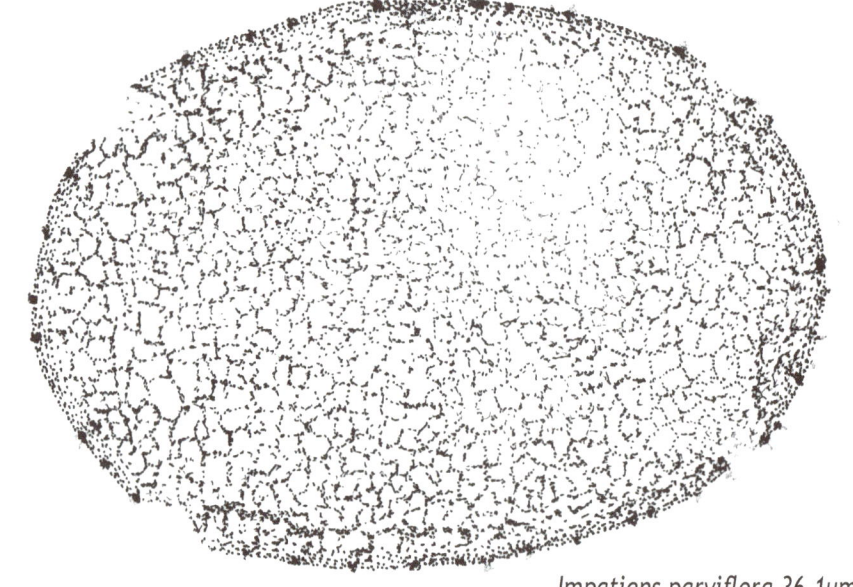

The impatiens family elongated shape and the indentation can be seen at each corner of the grain.

Impatiens parviflora 36.1µm
Balsaminaceae

Red Bartsia

A low growing, much branched semi parasitic plant found on poor soils. These pollen grains have six furrows three of which are higher than the alternate three in between.

Photo © Janet Morris

Odontites verna 26.6µm
Scrophulariaceae

Bay laurel

An evergreen shrub which although slow growing, can become quite large. It is dioecious, each tree bearing either male or female flowers.

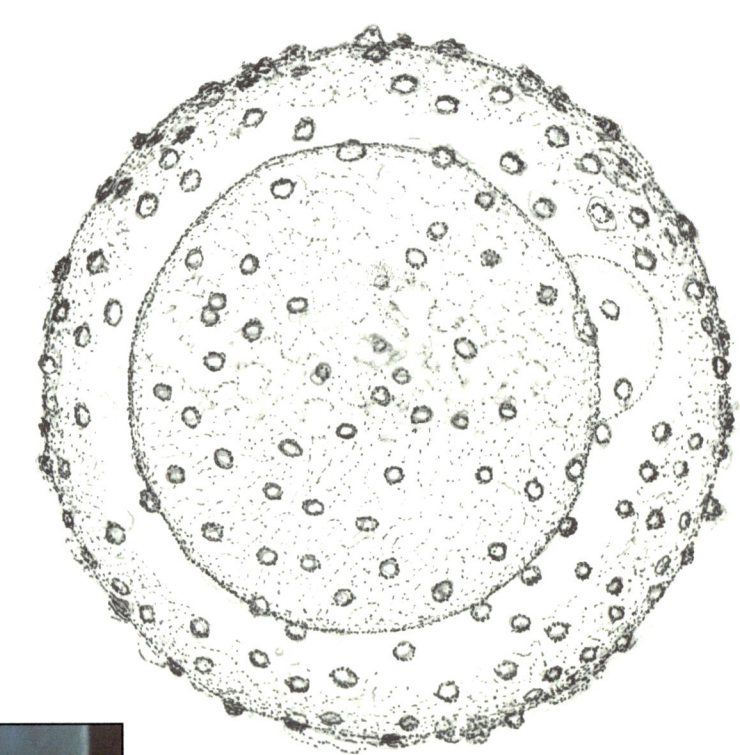

Laurus nobilis 51.2μm
Lauraceae

Beech

Fagus sylvatica 48.6μm
Fagaceae

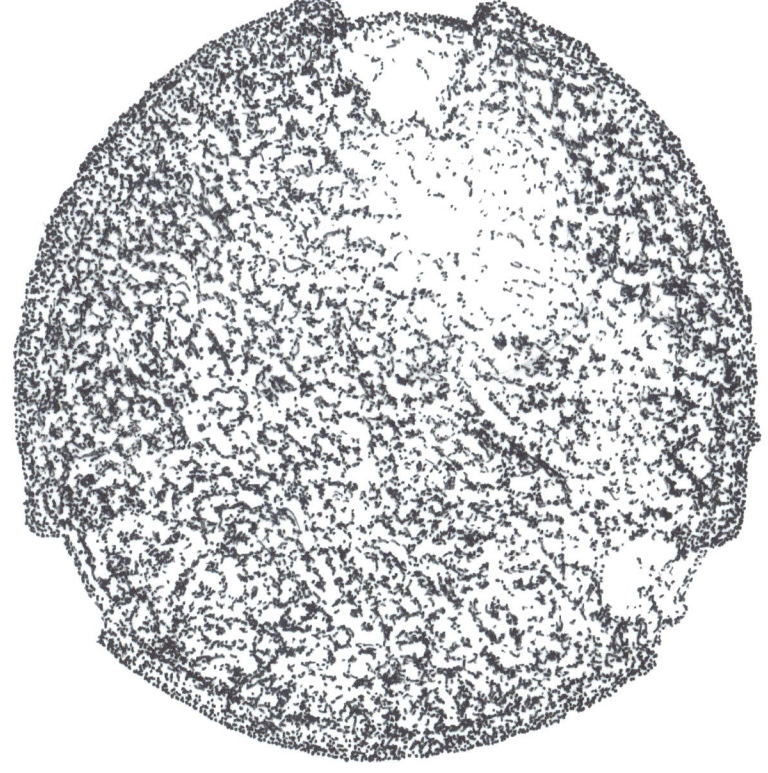

Beech is monoecious, and grows into a large tree. It flowers in April and May. The tassel-like male catkins hang from long stalks at the end of twigs, while female flowers grow in pairs, surrounded by a cup.
The pollen grain is quite large for wind pollinated dispersal. It has three long narrow furrows and the surface is tectate, ie with its exine sculpturing fused.

Peach leaved bellflower

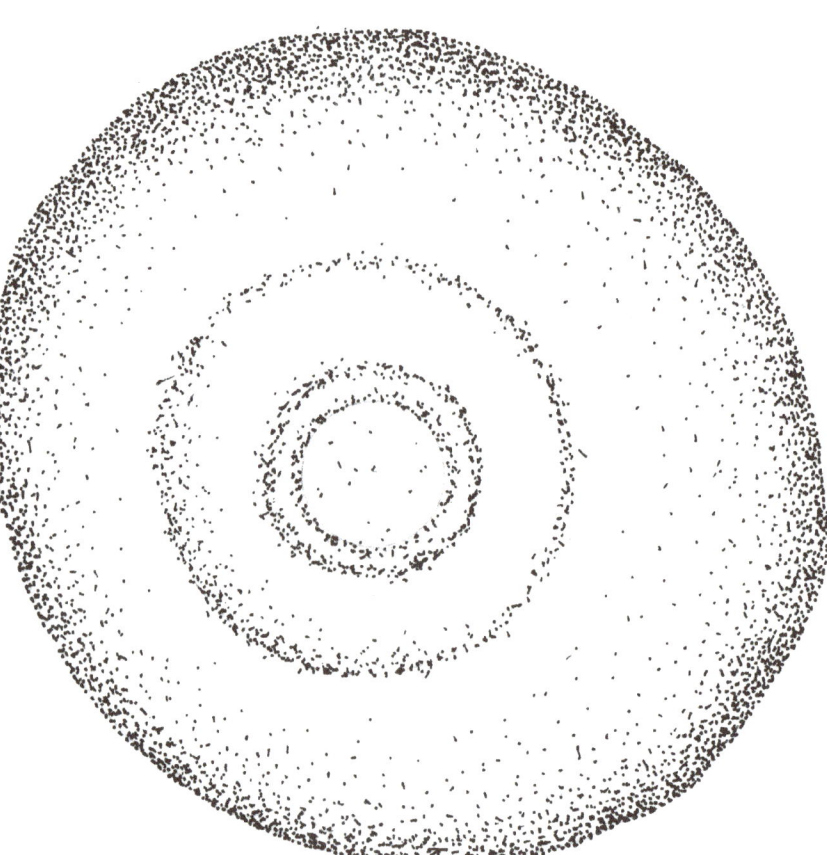

Campanula persicifolia 34.7µm
Campanulaceae

A summer flowering herbaceous perennial plant, the pollen grain, shown in equatorial view, has three or four pores spaced evenly around its axis.

Photos © Sally Dunn

Betony

A grassland herb visited by bees and other insects for its nectar. The pollen grain, shown in polar view, shows three wide furrows.

Photo © Sally Dunn

Stachys officinalis 24.0μm
Lamiaceae

Bindweed

Strong, smothering, hedge climber which can be a perennial nuisance, spreading via a very persistent, resilient root system. Individual flowers last for one day. The pollen is dehisced in the morning, though the flower stays open through until the evening when it is visited for its nectar by late flying insects such as hawk moths.
The large pollen grain is finely dotted and covered in pores at regular intervals over the whole sphere.

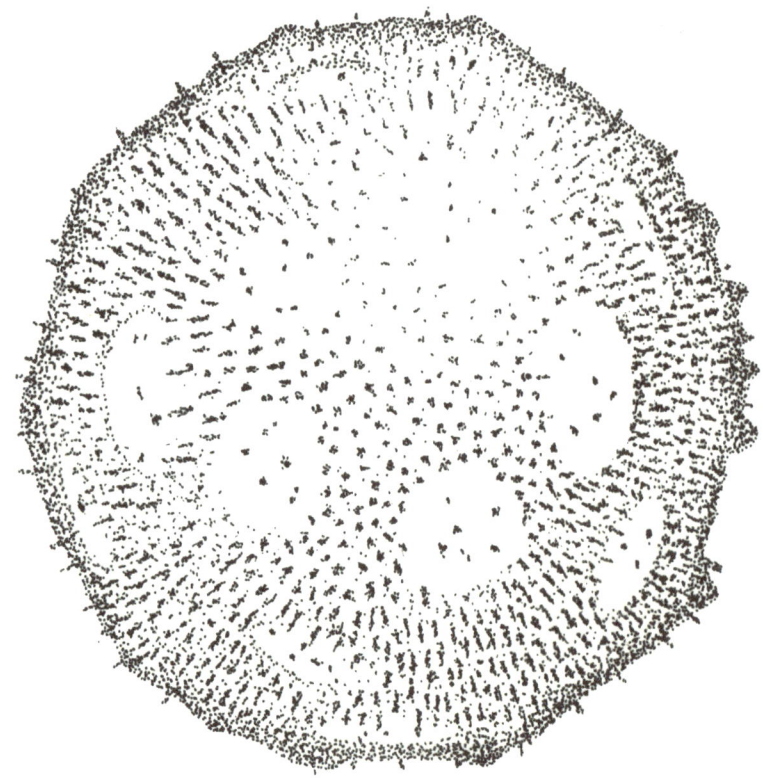

Calystegia silvatica 90.0µm
Convolvulaceae

Black bindweed

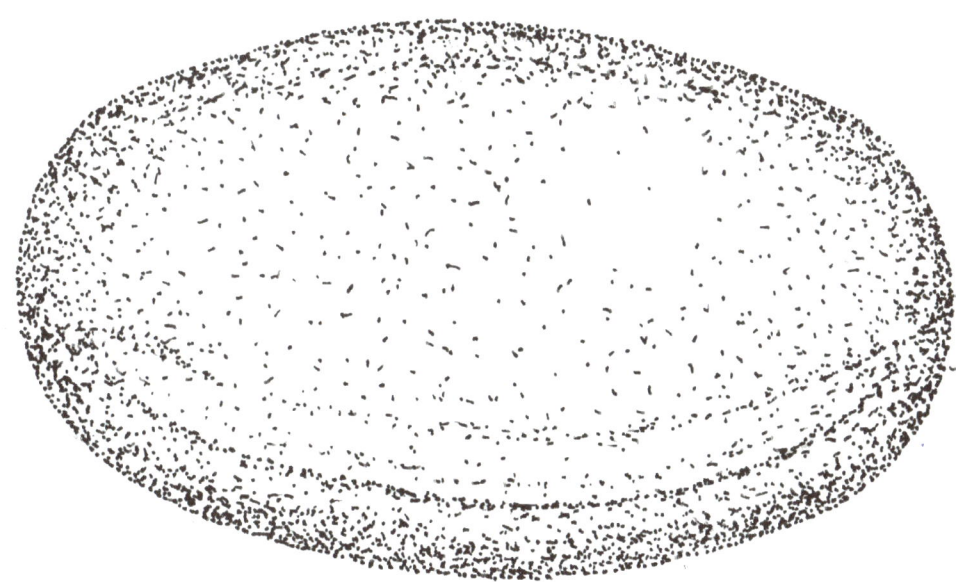

Fallopia convolvulus 22.3µm (Bilderdykia convolvulus)
Polygonaceae

An herbaceous vine which twists around other plants, also known as wild buckwheat. It has small heart shaped leaves and unobtrusive flowers. Although the seeds are edible the yield is too low for a commercial crop.

Bird's foot trefoil

Perennial herbaceous wild flower visited mostly by bumble bees. The pollen grain is slightly elongated. It has three forrows each with a pore at the centre.

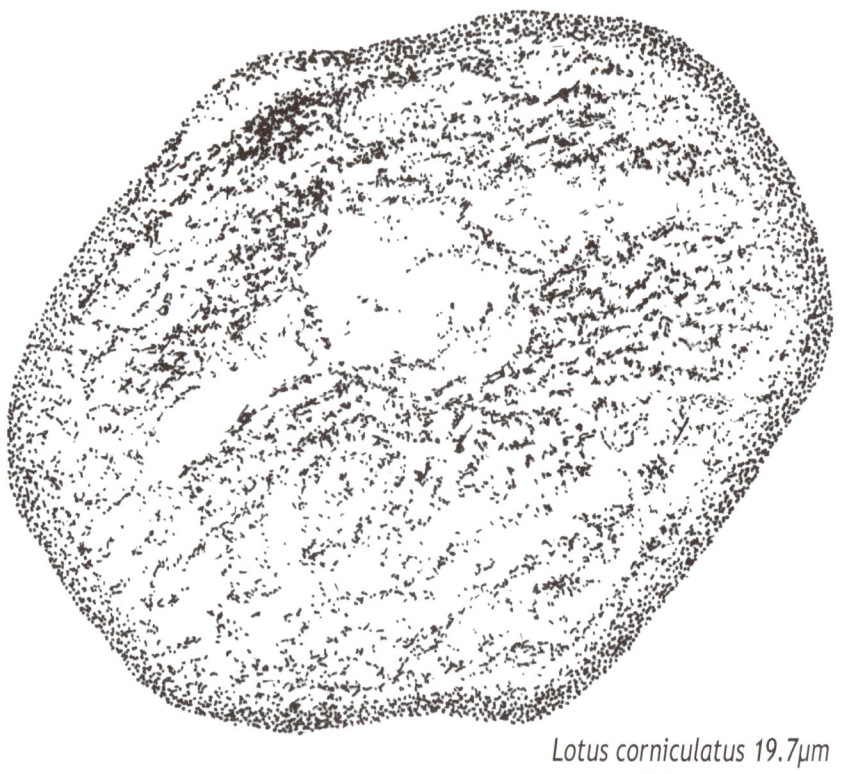

Lotus corniculatus 19.7µm
Fabaceae

Bittersweet

Photos © Sally Dunn

Members of the nightshade family range from quite deadly to food plants we know well such as tomatoes, potatoes, peppers and aubergines. In common with others in the family, pollen is released by the vibration of visitors such as a bumble bee.
The pollen grain is very small, tricolpate, with long furrows and a smooth surface.

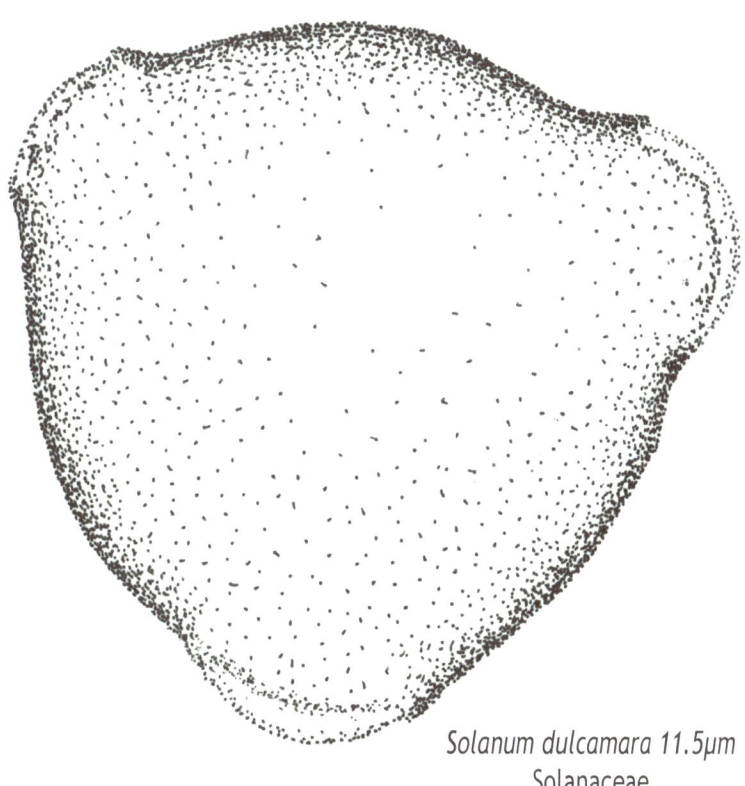

Solanum dulcamara 11.5µm
Solanaceae

Blackthorn

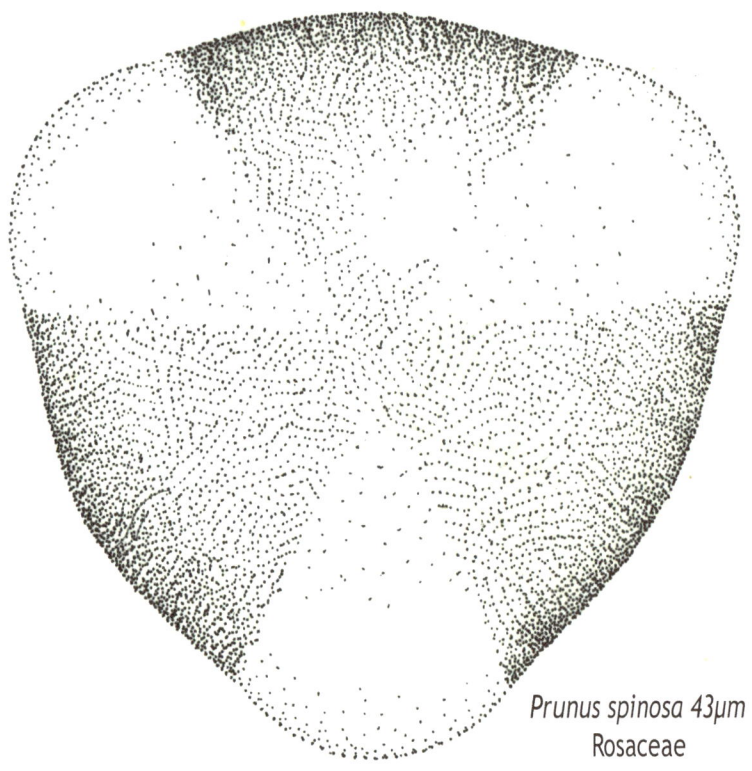

Shrub, named from the colour of its bark, with sharp spines and flowers appearing before the leaves. It produces dark blue berries known as sloes in the autumn.
The pollen has three wide furrows and a striated surface pattern.

Prunus spinosa 43µm
Rosaceae

Bluebell

A woodland perennial bulb, the English Bluebell has a drooping flower with the 'bells' pointing in the same direction from the stem. It hybridises readily with the introduced Spanish bluebell.
Each has a pollen grain with one furrow.

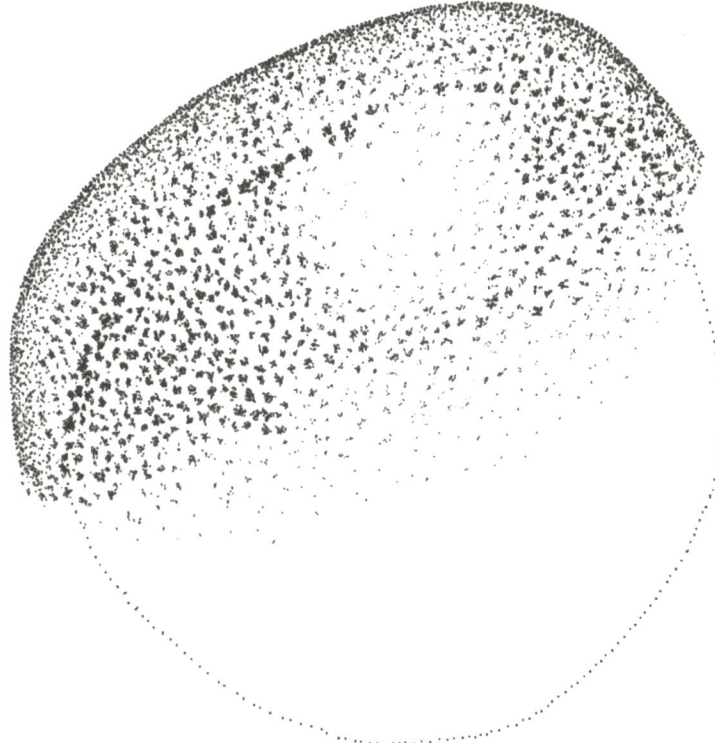

Hyacinthoides non-scripta (formerly Endymion non-scriptus) 62.1μm
Asparagaceae

Borage

Edible, annual herb with attractive star shaped flowers. High demand for the borage oil, which is rich in gamma-linoleic acid, is resulting in an increase in the spring crop being grown in the UK. The ten furrows make the pollen grain easy to recognise.
The drawing shows a polar view, the grain being elongated as is typical of the borage family.

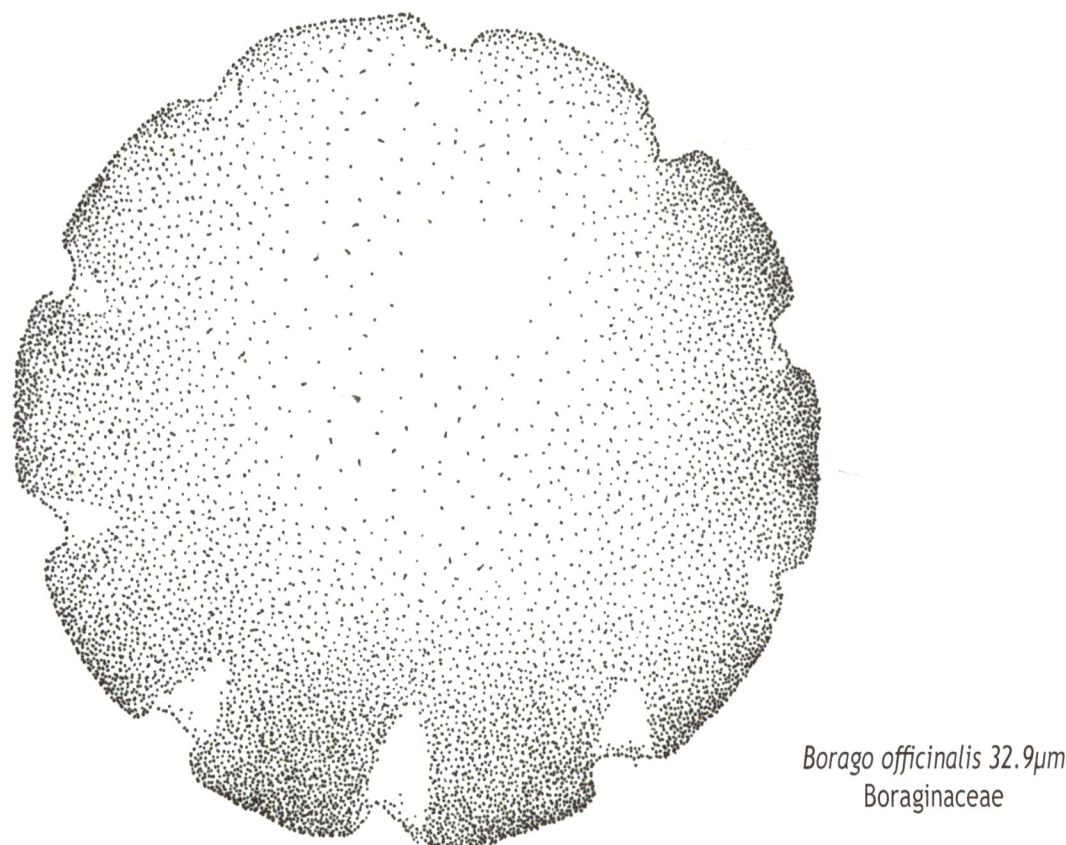

Borago officinalis 32.9µm
Boraginaceae

Box

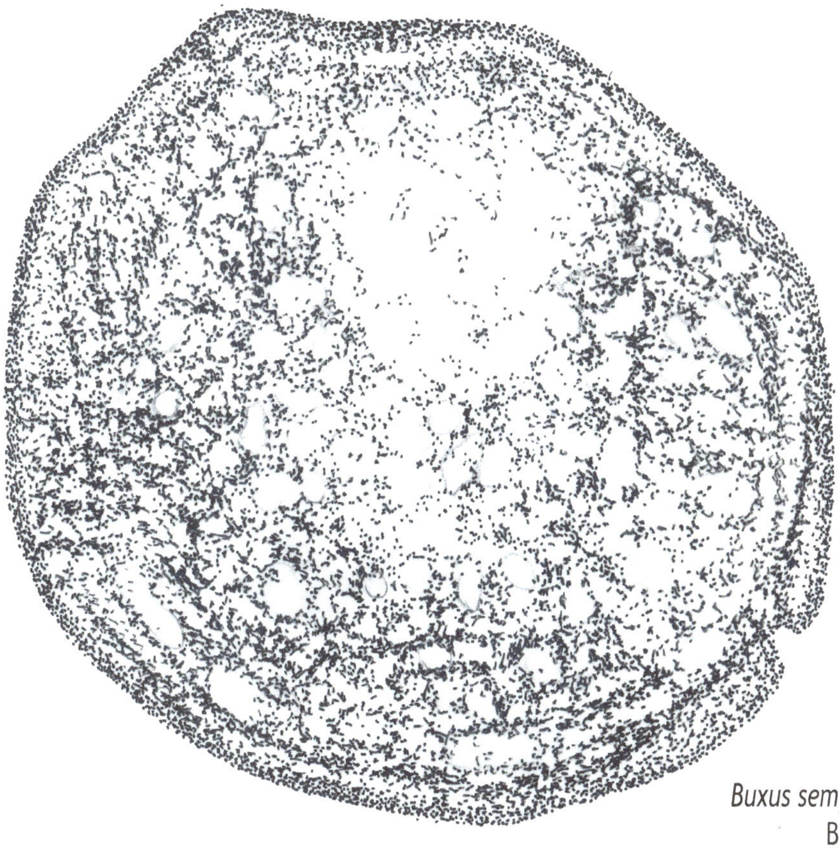

Buxus sempervirens 26.4μm
Buxaceae

Compact, slow growing, hardy shrub often used for hedging. Very low growing varieties are often used for edging in formal gardens. The pollen grain has numerous pores but no furrows.

Bramble (Blackberry)

Prickly, unruly, tangly bush, providing food and cover for many insects (including hoverflies as shown below) and other creatures, both at flowering and fruiting time. This plant is notoriously variable and the pollen grains are no exception.

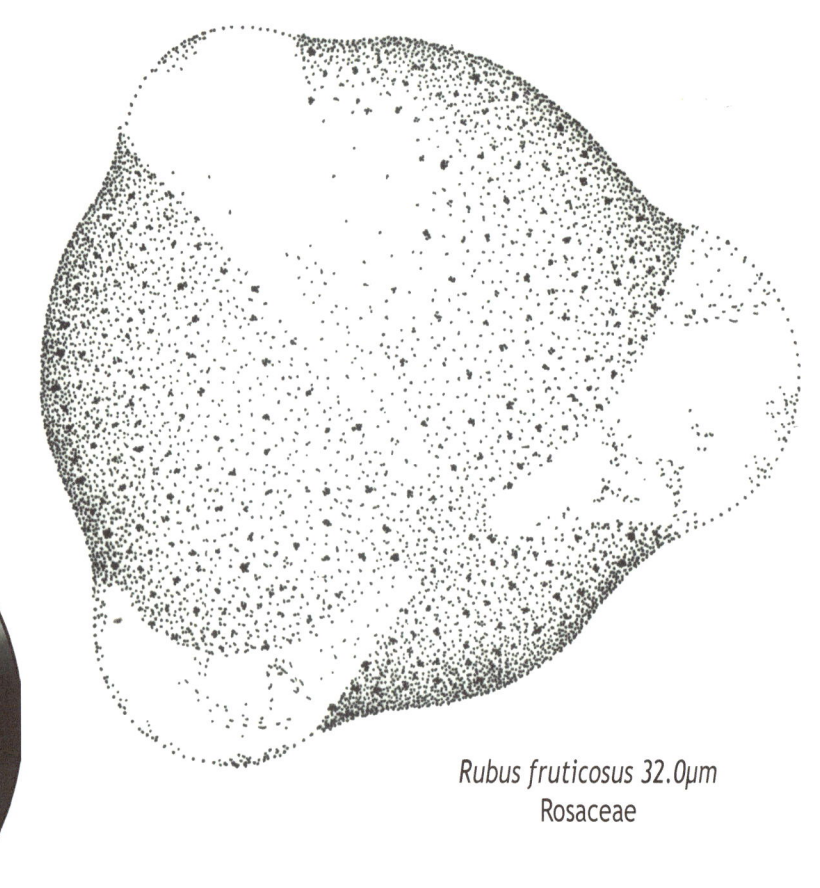

Rubus fruticosus 32.0μm
Rosaceae

Broom

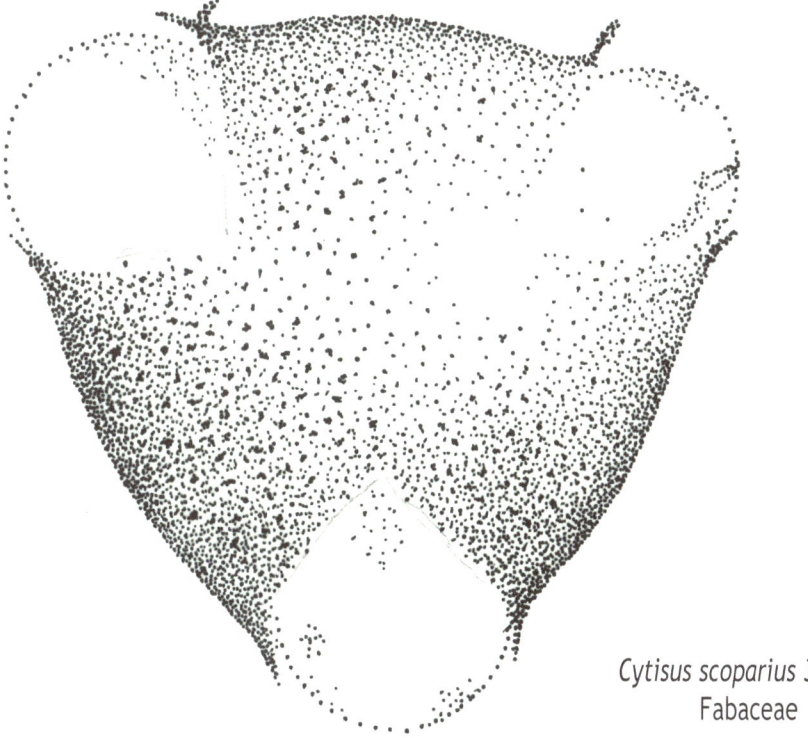

A perennial leguminous shrub, the flowers are fused shut and require forced "tripping" by pollinating honey or bumble bees.

Cytisus scoparius 31.3μm
Fabaceae

Bryony (white)

White bryony, a vigorous hedgerow climber, distinguishable from the unrelated black bryony by its 5-lobed leaves and tendrils, can pull down branches with its weight. Bryony is dioecious, having male and female flowers on different plants. The male flowers have five stamens, covered in pollen at the tips. One of the solitary ground-nesting bees, *Andrena florea*, will only collect pollen from white bryony.

Bryonia dioica 42.3μm
Cucurbitaceae

Buckthorn (purging buckthorn)

Purging buckthorn, often referrred to just as buckthorn, is a shrubby, spiny, small tree. Its yellow green, four petalled insect pollinated male and female flowers are found on separate trees.
The pollen grain is of a similar size and appearance to yew pollen.

Rhamnus cathartica 25.2μm
Rhamnaceae

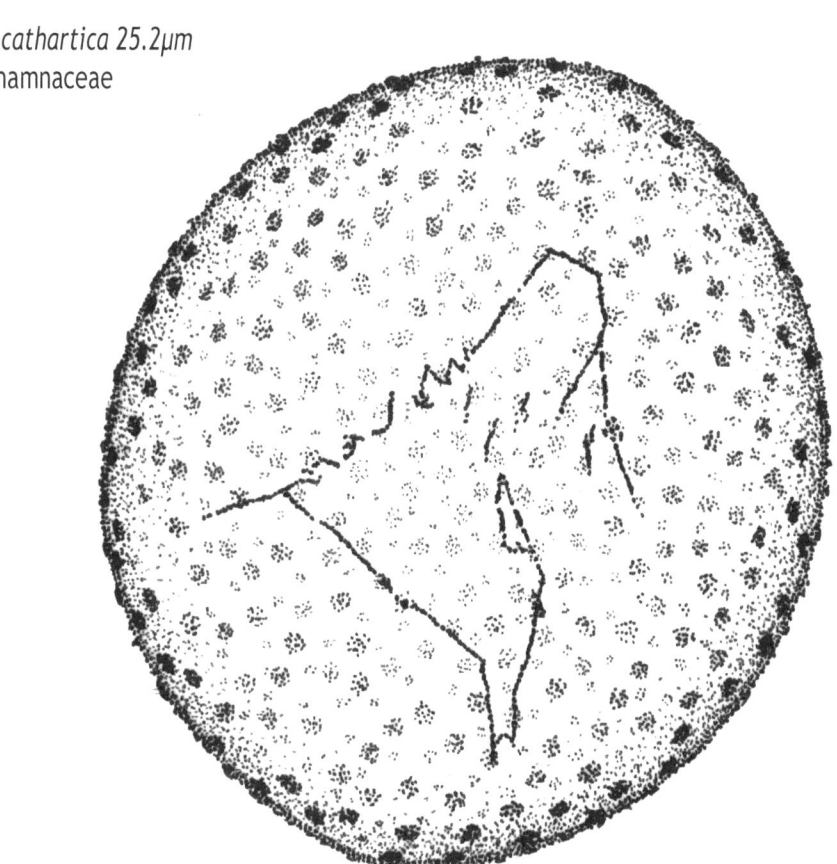

Buddleia

The 'butterfly bush' is visited by many insects for its flowers, rich in nectar. The anthers are halfway down the corolla tube. The pollen grains have four furrows.

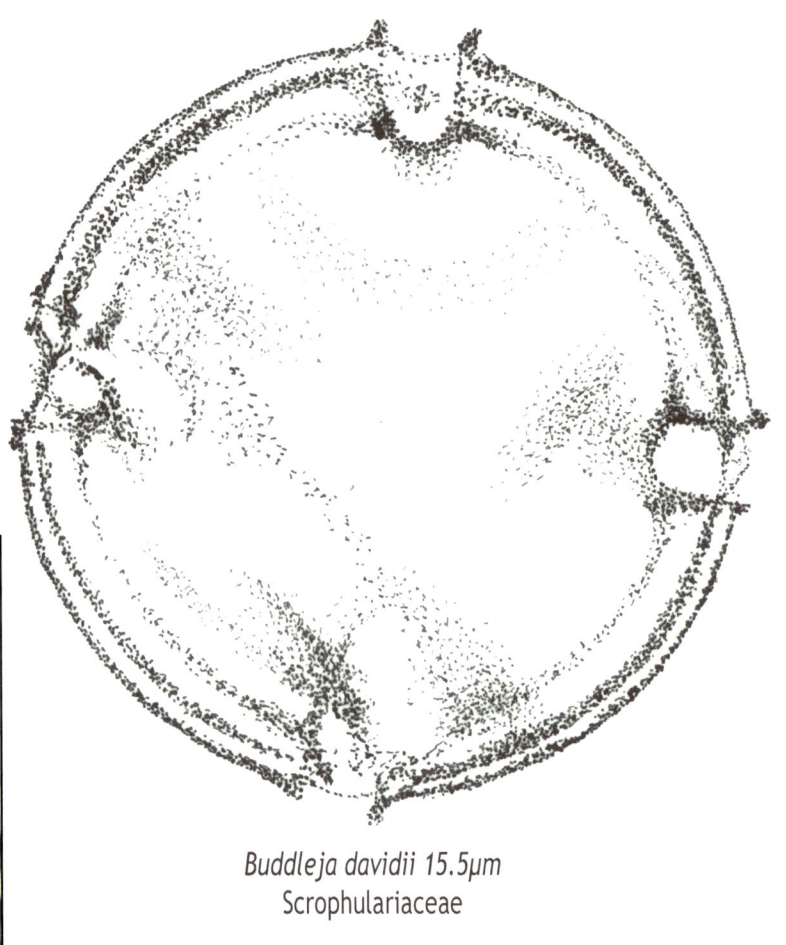

Buddleja davidii 15.5μm
Scrophulariaceae

Burnet saxifrage

The two angles of pollen grain demonstrate the three furrows each with a pore in the centre of the furrow. The long pollen grain is typical of umbellifers.

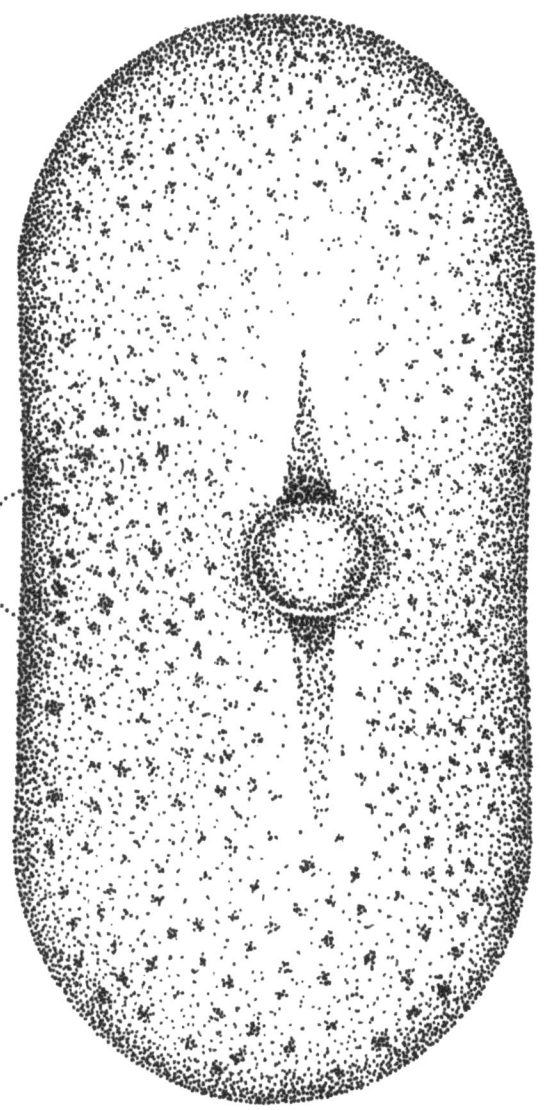

Pimpinella saxifraga 38.6µm
Apiaceae (Umbelliferae)

Busy Lizzie

Impatiens walleriana 45.1µm
Balsaminaceae

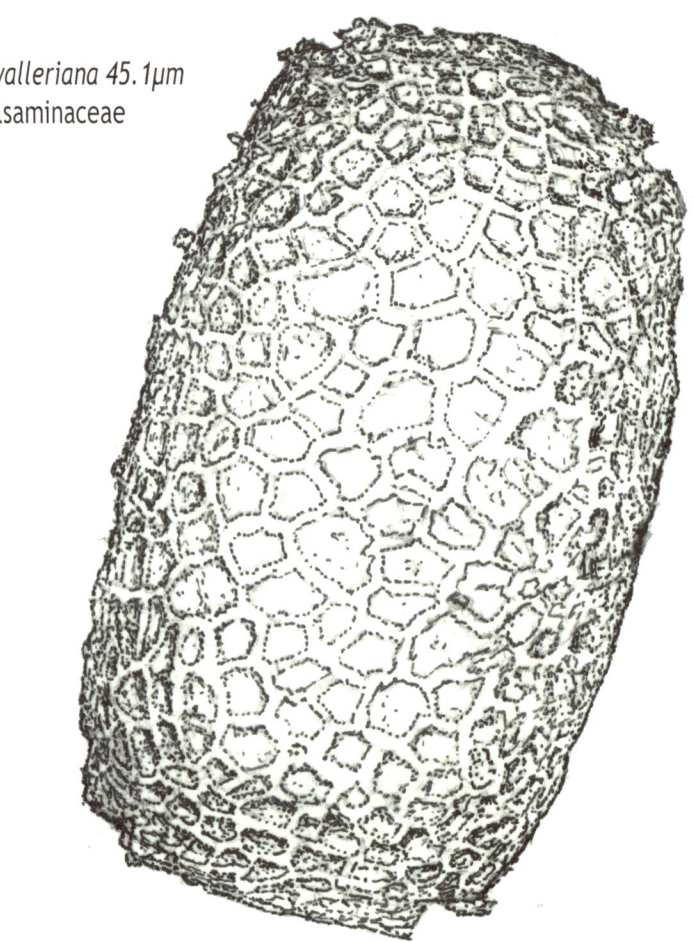

The oblong shape and netting with raised 'walls' surrounding small areas on the pollen grain surface is peculiar to the impatiens family.

Meadow buttercup

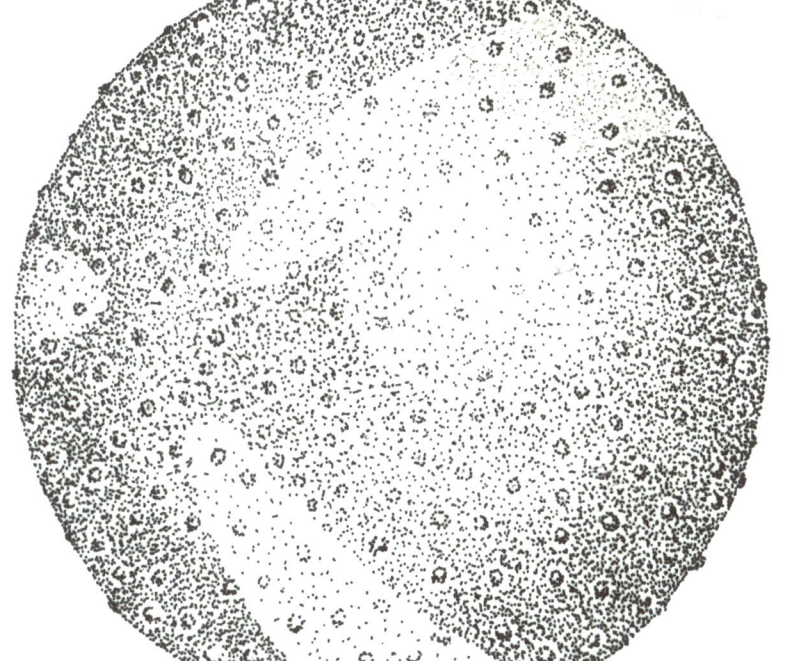

An attractive but invasive wild flower. Buttercups have a spherical pollen grain with a typical configuration of three furrows.

Ranunculus acris 37.7µm
Ranunculaceae

Wild cabbage

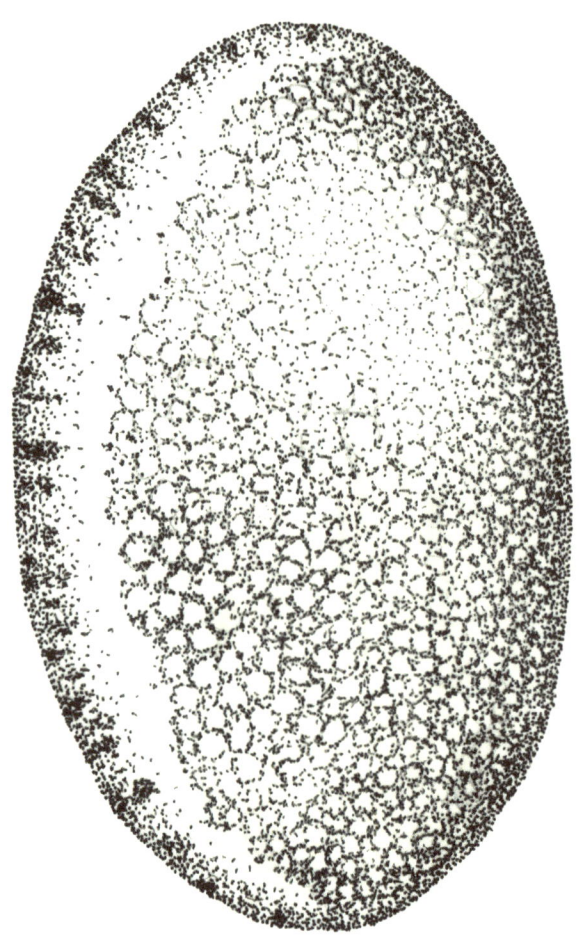

Brassica oleraceae 28.6µm
Brassicaceae (Cruciferae)

The pollen grain shape is prolate or elongate so lies on its side with the view under the microscope always an equatorial view.

Christmas cactus

Native to Brazil, a characteristic of this popular house plant is that the many stamens are arranged in two series: the inner stamens form a ring around the style; the outer stamens arise from the floral tube. The filaments of the stamens are white, the anthers and pollen being yellow. The style has six to eight lobes at its end and is dark red.

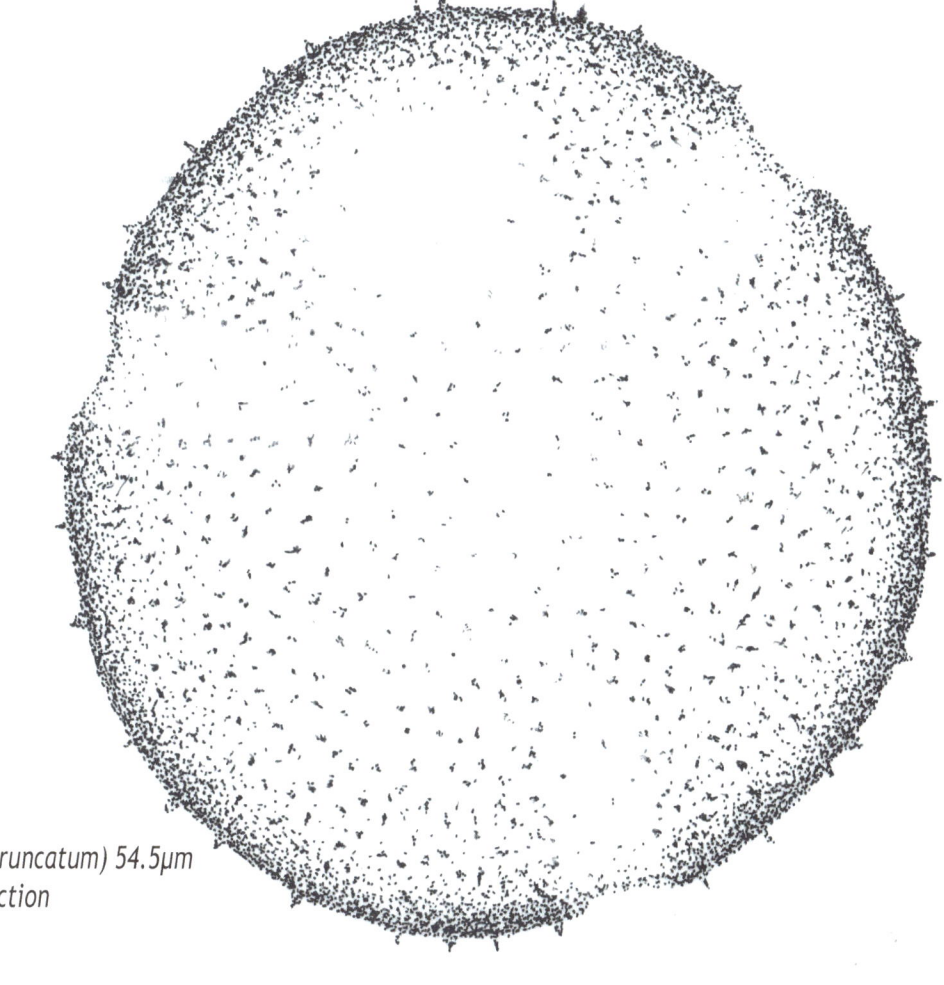

Schlumbergera truncata (Epiphyllum truncatum) 54.5µm
Slide from Quekett collection
Cactaceae

Camellia japonica

From the same ericaceous family as *Camellia sinensis*, the tea plant, this evergreen shrub has leathery leaves and spectacular flowers. The pollen has a reticulated surface and decoration on the surface of the three furrows.

Camellia japonica 52.0µm
Theaceae

Bladder campion

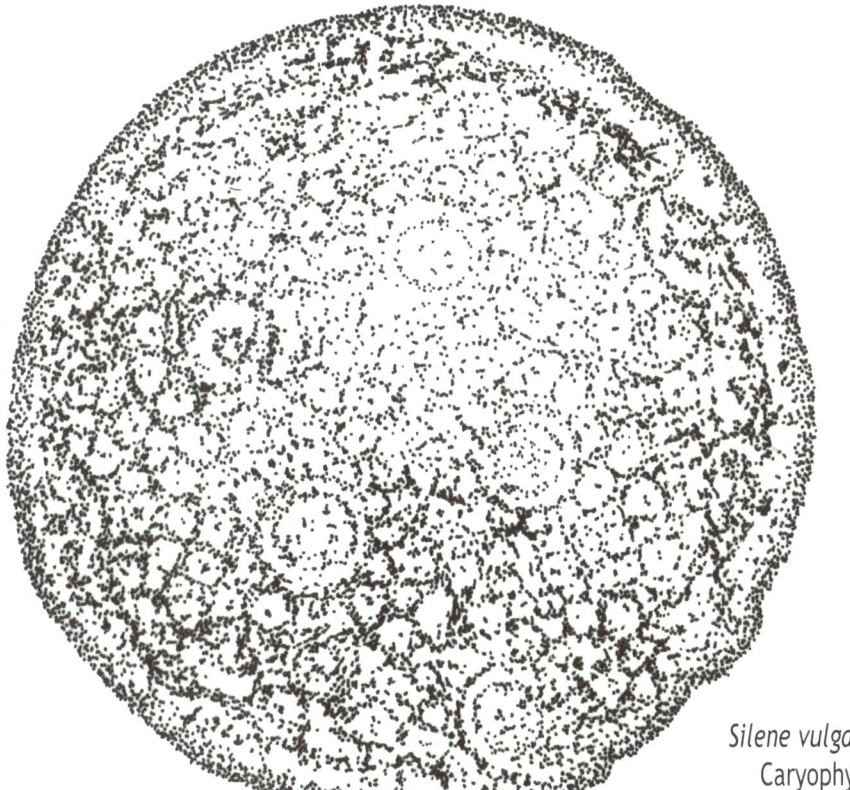

These flowers have a pinkish inflated calyx with a net pattern.
The pollen grain surface is covered with bead like sculpturing elements (gemmae) and multiple pores.

Silene vulgaris 37.0µm
Caryophyllaceae

Photo © Sally Dunn

White campion

White campion is a dioecious flowering plant, distinguishable from bladder campion having ridges on the enveloping calyx.
The pollen grains are similar in size and appearance, having the bead like surface interspersed with pores.

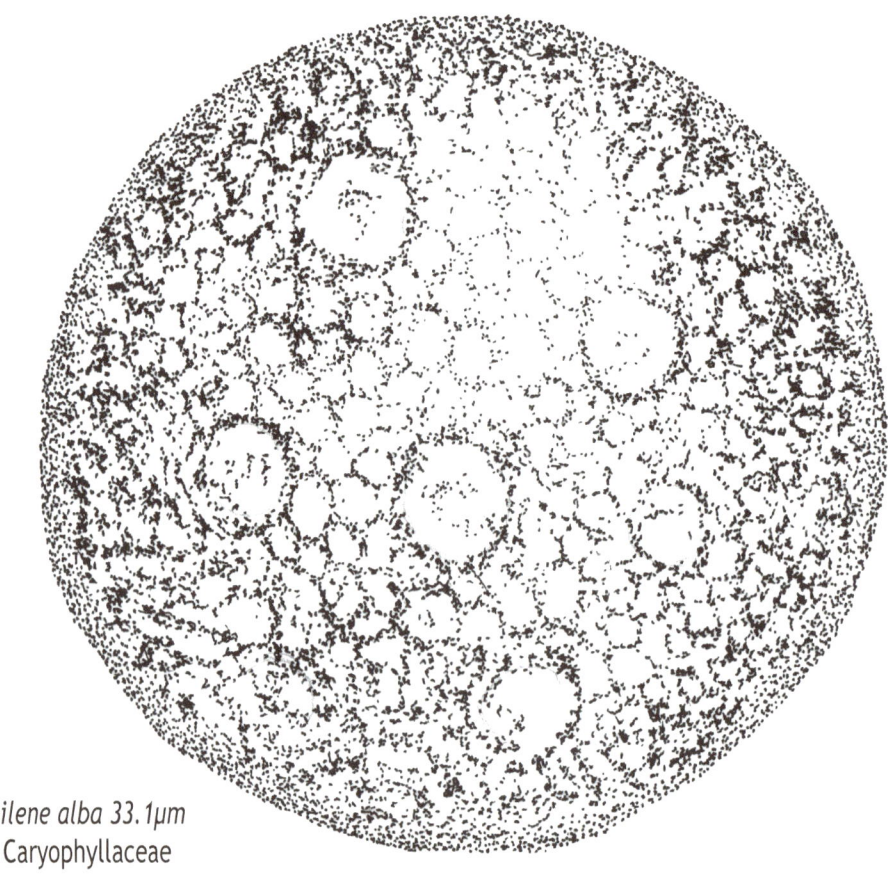

Silene alba 33.1µm
Caryophyllaceae

Male flowers (same form as for these pink campion flowers left) are on different plants from female flowers (right)

Candytuft

Photo © Sally Dunn

The pollen grain shows a netting pattern on the surface and the illustration hints at the three furrows.

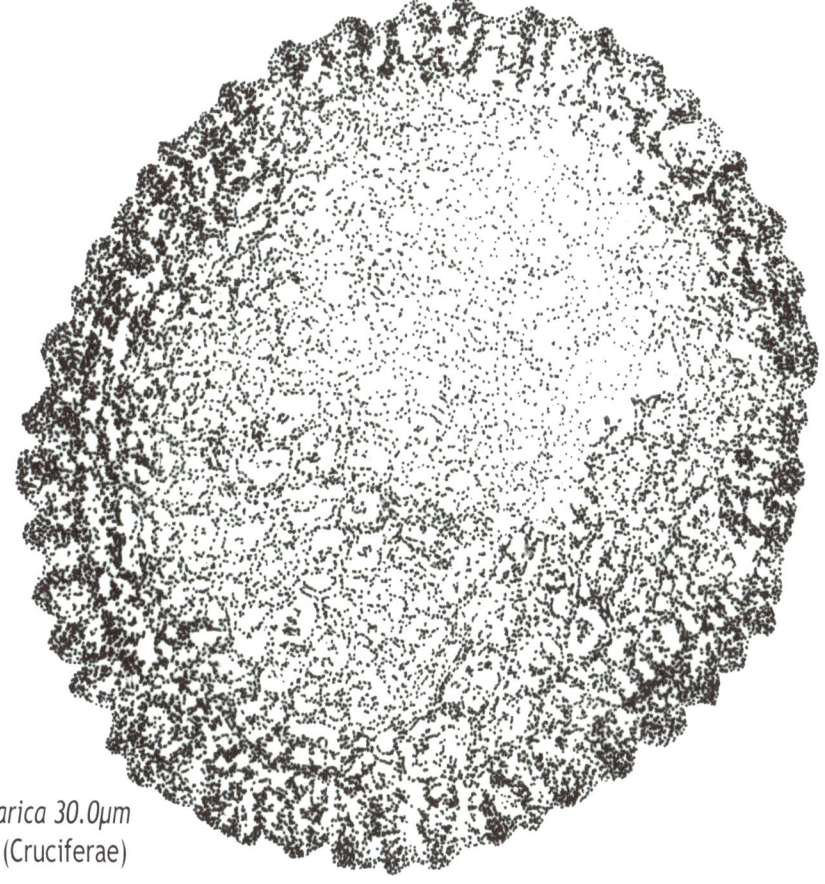

Iberis gibraltarica 30.0µm
Brassicaceae (Cruciferae)

Ceanothus

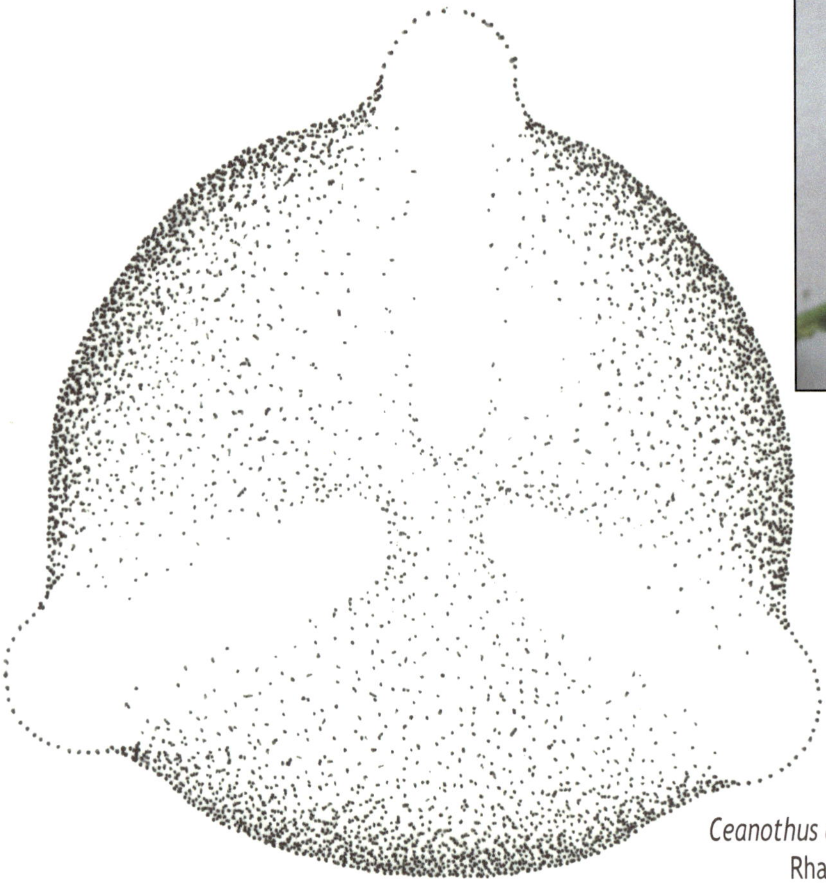

Spring flowering shrub, attractive to insects. The pollen gain has a small polar field.

Ceanothus arboreus 26.1μm
Rhamnaceae

Greater celandine

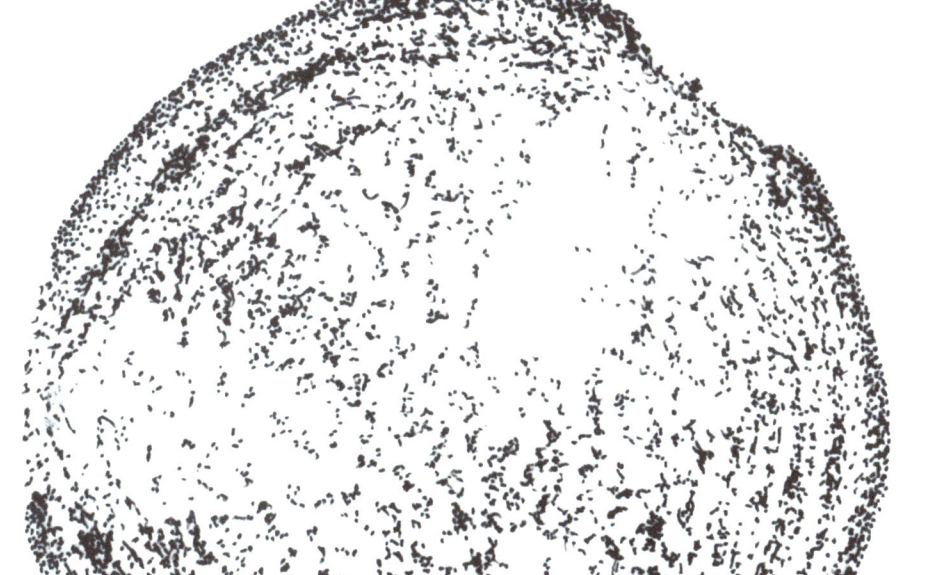

Spheroidal pollen grain with three furrows. The grain has a finely reticulated surface and looser pattern of decoration across the furrows.

Chelidonium majus 22.9μm
Papaveraceae

Lesser celandine

This pollen grain has a dotted surface. Electron micrographs show a closer pattern of dotting on the furrows.

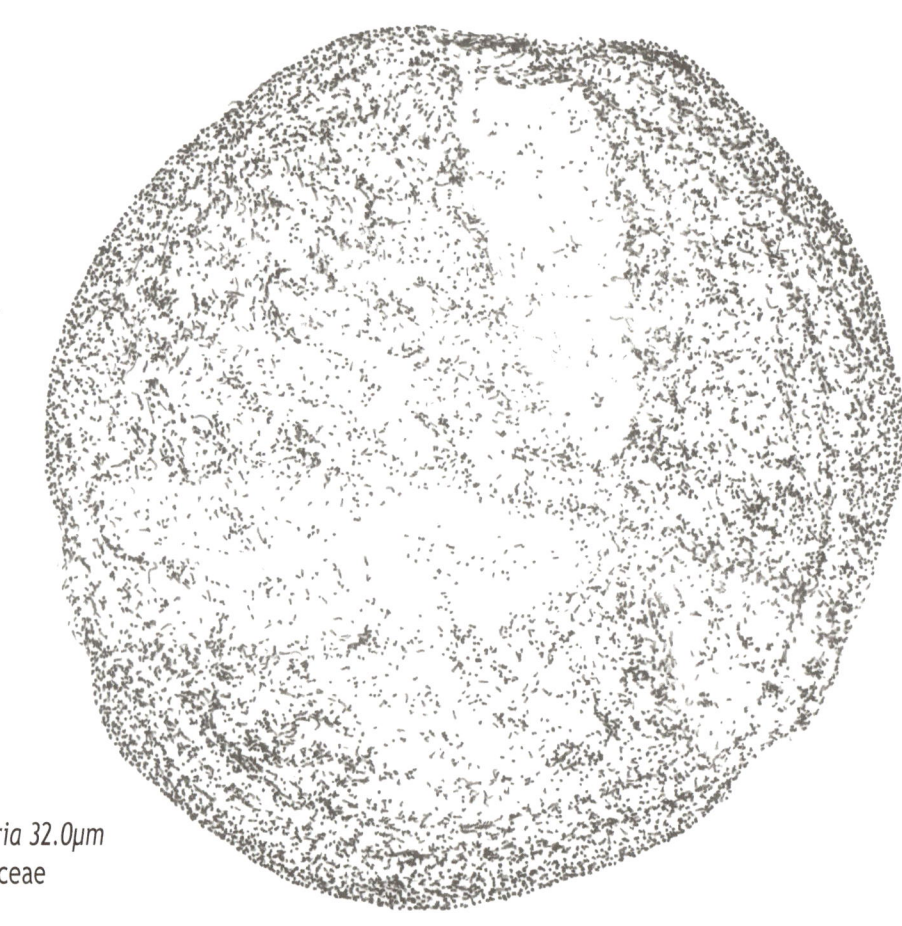

Ranunculus ficaria 32.0μm
Ranunculaceae

Flowering cherry

Prunus sp. 38.9μm
Rosaceae

Polar view of the pollen grain showing the large pores. Not always easy to see using a light microscope, the surface of the *Prunus* pollen grain is typically striated, sometimes quite straight lines, sometimes quite swirly.

Chickweed

Chickweed, a weed of cultivation, will cover a large area with its foliage. The tiny white flowers frequently pass unseen. The pollen grains have about eighteen pores in a global arrangement but no furrows. The grains also seem to be peppered all over with tiny dots, particles attached to the surface. Those that occur on the pores tend to collect on the pore centres. Those shown on the illustration are borne on the crown of each dome of intine bulging through the pore aperture and hold the cluster of particles aloft, well above the surface level of exine surrounding the pores.

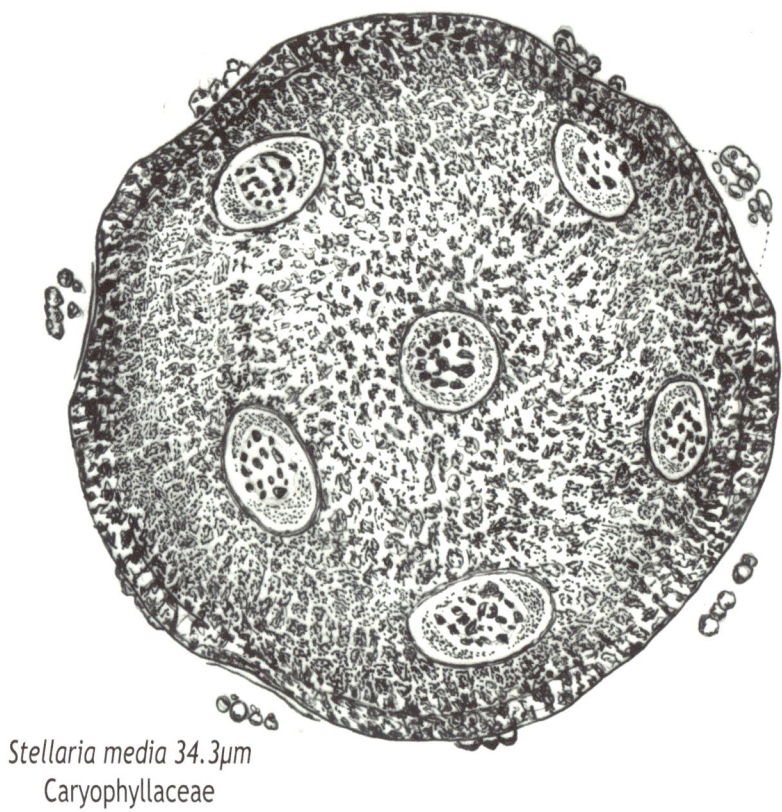

Stellaria media 34.3µm
Caryophyllaceae

Photo © Sally Dunn

Chicory

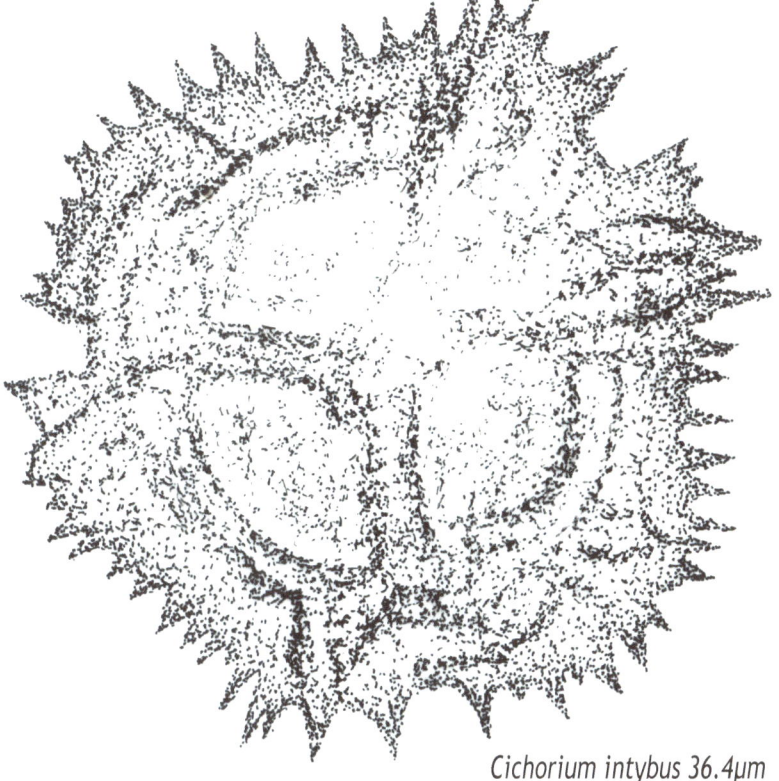

Cichorium intybus 36.4µm
Asteraceae (Compositae)

A pretty, usually blue, wild flower seen on roadsides. Cultivated forms of chicory have many culinary uses.
The pollen grain has a fascinating sculpture. It has apertures with large pores within a pattern of regular oval or pentagonal valleys surrounded by spiked ridges.

Chives

A well known herb in the onion family having the usual boat shaped, single furrowed pollen grain.

Allium schoenoprasum 27.3µm
Alliaceae

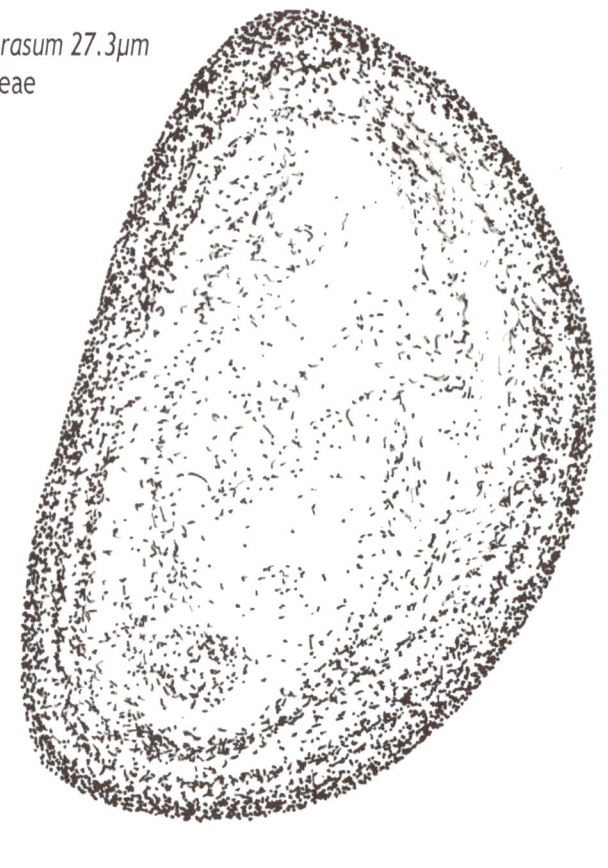

Common cleaver

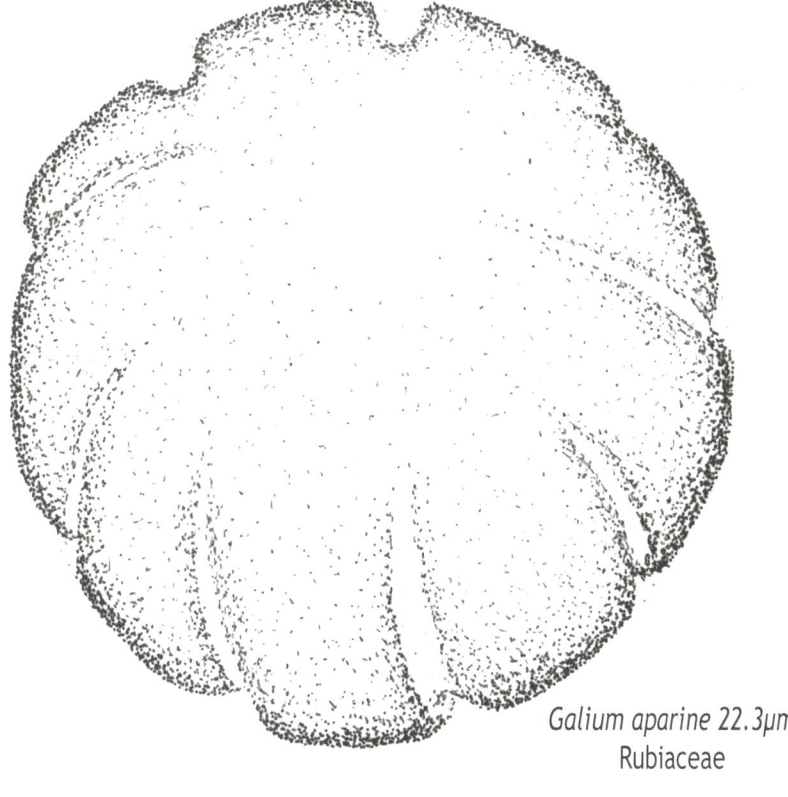

A slender, sticky, clinging climber with tiny flowers and several common names including goosegrass. Grains bear six to nine evenly distributed furrows.

Galium aparine 22.3μm
Rubiaceae

Photo © Sally Dunn

Clover (red)

Red clover is a meadow plant, often growing taller than white clover. Clovers, like other members of the bean family, are useful nitrogen fixing plants.
The red clover pollen grain has a netted surface. Grains of both red and white clover have three furrows.

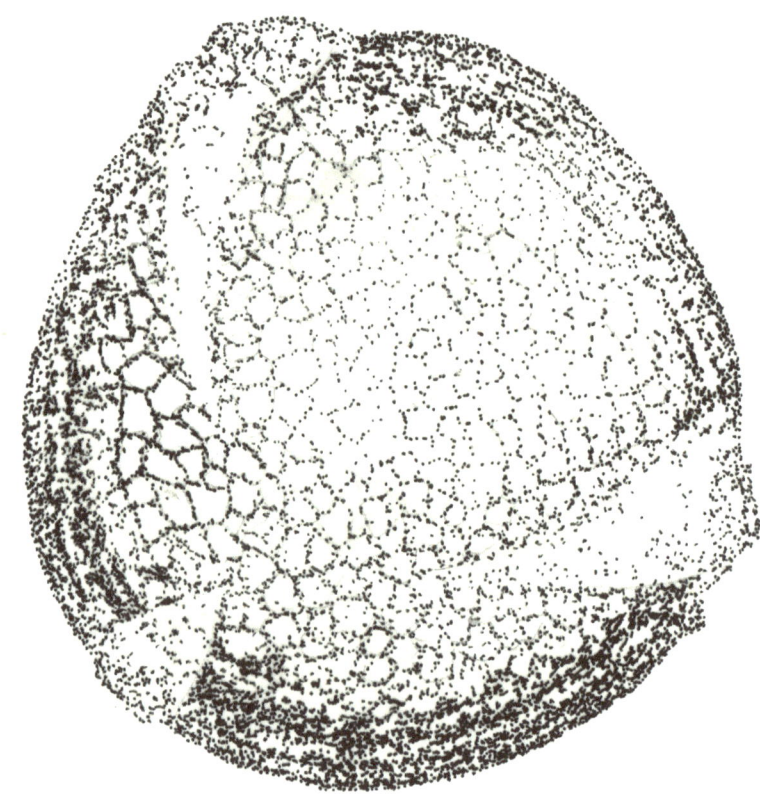

Trifolium pratense 35.0μm
Fabaceae

Clover (white)

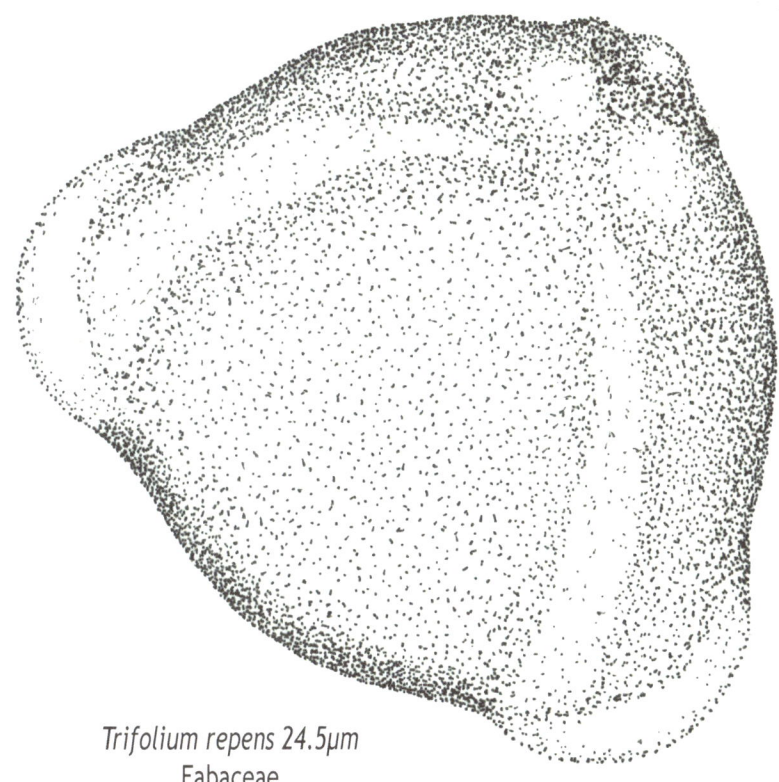

Trifolium repens 24.5µm
Fabaceae

White clover usually forms a low growing, spreading, perennial ground cover among grass. It provides extensive forage not only for insects, but also farm livestock. The surface of the grain is dotted.

Photo © Gerry Collins

Cocksfoot

Photo © Sally Dunn

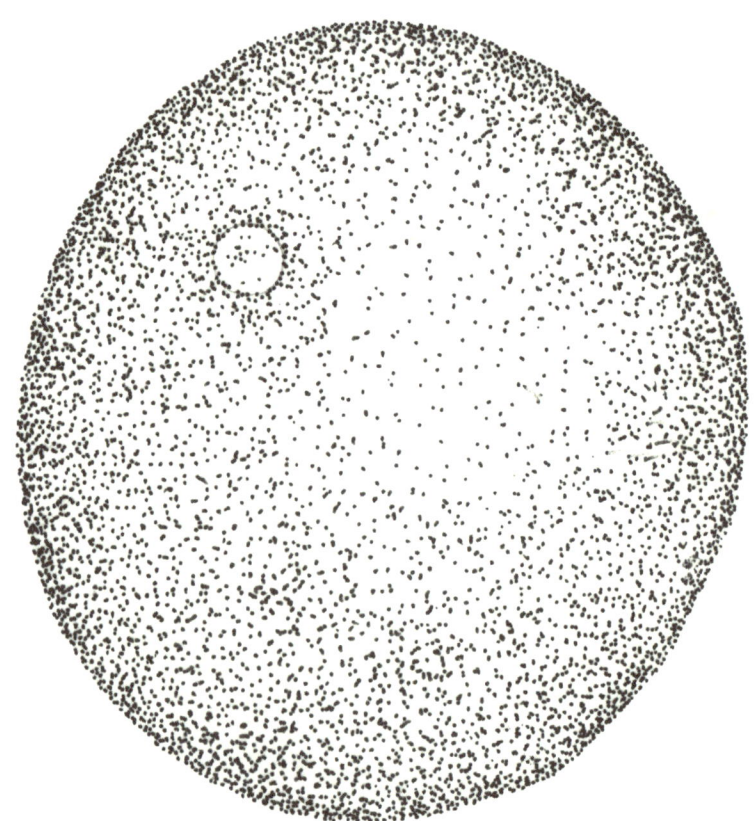

Dactylis glomerata 36.0µm
Poaceae (Gramineae)

Photo © Janet Morris

These grass flowers are quite distinctive. The top photo is of the seed head, the bottom photo of the flower showing cream coloured stamens, but the stamens are very often red.
Cocksfoot grass is wind pollinated. The pollen grain typically has just one pore.

Coleus canina (scaredy cat plant)

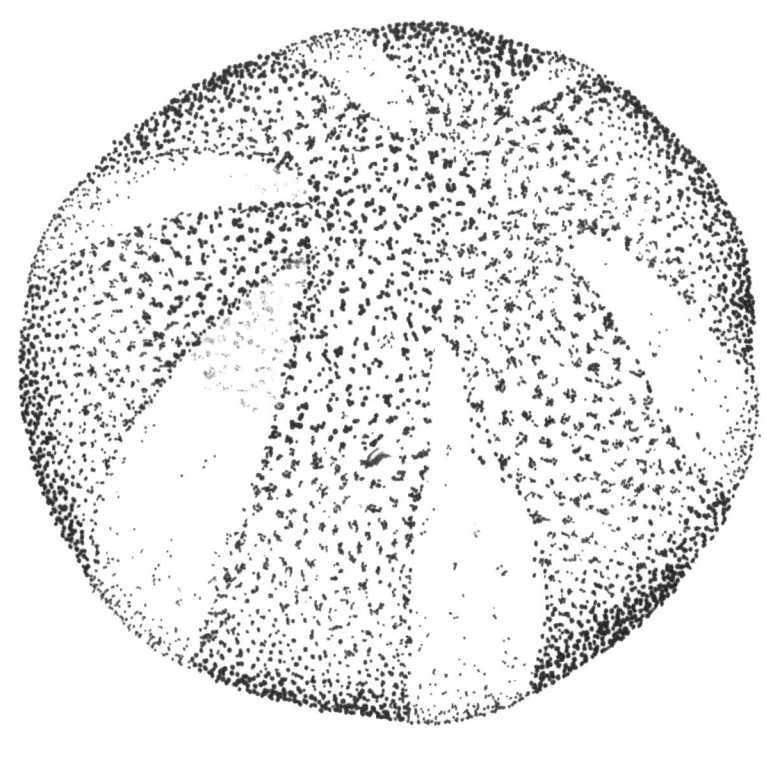

A member of the mint family, the pollen grain has six furrows; the exine shows granulation at the poles but a net pattern elsewhere.

Plectranthus caninus 49.0µm
Lamiaceae

Coltsfoot

Perennial, with edible flowers which appear in spring before any leaves. The pollen grains have three furrows and a spiky exine covering.

Tussilago farfara 46.0µm
Asteraceae (Compositae)

Comfrey

Photo © Sally Dunn

Native to Europe, varieties of comfrey have been developed as an organic fertiliser. Comfrey pollen grains have ten pores spaced around the equator. The pollen grain shape is prolate or elongated so lies on its side with the view under the microscope always an equatorial view.

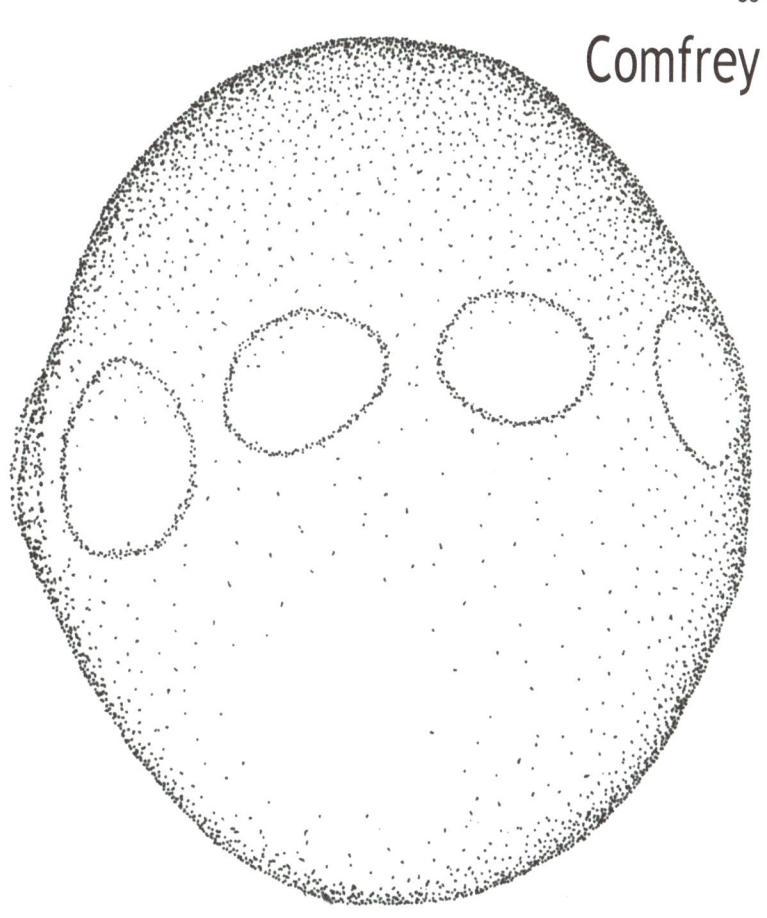

Symphytum officinale 26.5µm
Boraginaceae

Coral bells

Garden hybrids are developed and grown for their foliage, the flowers being comparatively insignificant but nevertheless attractive to bees.

Heuchera sanguinea 31.6μm
Saxifragaceae

Photo © Sally Dunn

Coriander

Both the leaves and seeds are used as a cullinary herb. The flowers, attractive to insects, are set in umbels with the outer petals longer than those around the centre.
The pollen grain is elongated with three pores spaced equally around the equator.

Coriandrum sativum 28.2µm
Apiaceae (Umbelliferae)

Cornflower

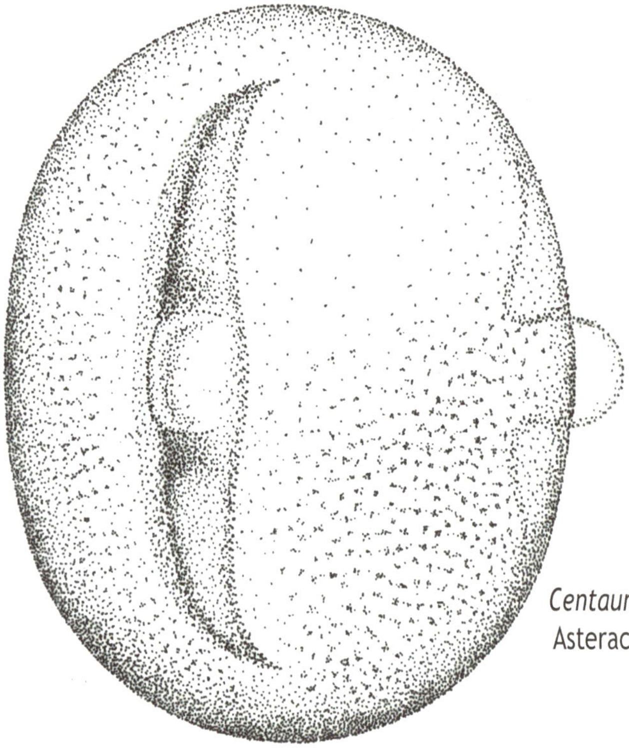

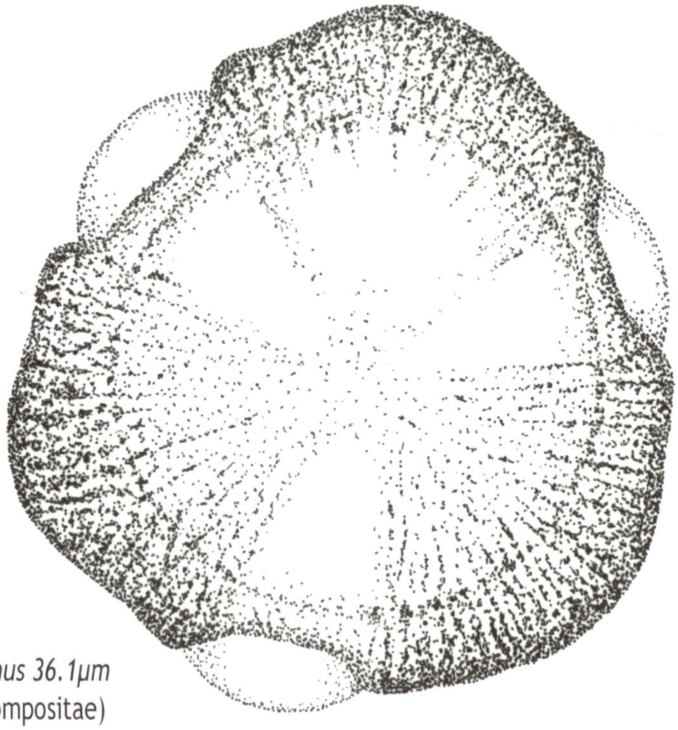

Centaurea cyanus 36.1μm
Asteraceae (Compositae)

One of the few pollens among the Asteraceae that have no spines.
Two equatorial views on the far left and polar view above.

Photo © Bill Fisher

Yellow corydalis

Corydalis lutea 44.0µm
Papaveraceae

Yellow corydalis pollen has a rare configuration with the six furrows lying along the edges of a regular tetrahedron. The tips come together in threes and actually meet. This occurs at four points on the sphere. The triangular segments of exine are often seen separated.

Cosmos

Annual in the daisy family, attractive to insects and long lasting as a cut flower. The pollen grain is covered in spikes at regular intervals, and has three pores, hardly visible until the grain is hydrated.

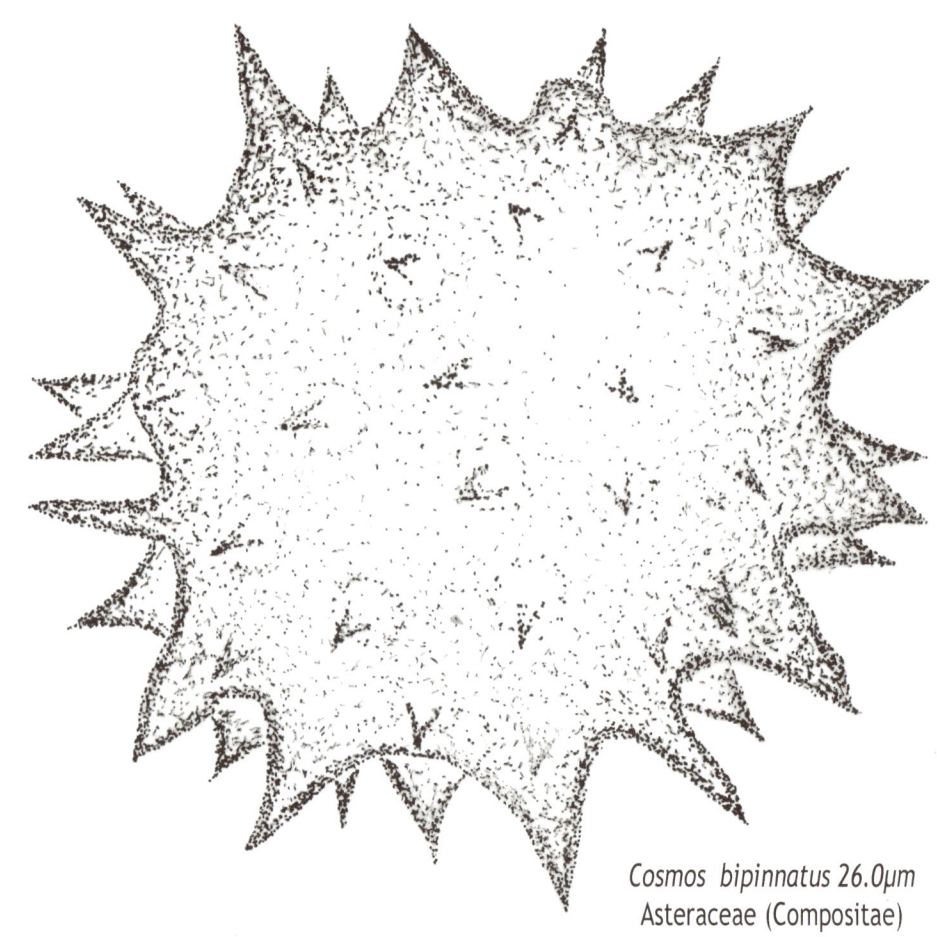

Cosmos bipinnatus 26.0μm
Asteraceae (Compositae)

Cow parsley

Three to four foot tall, spring flowering roadside, hedgerow plant with attractive frothy flower heads. All the umbellifer pollens are elongated. One can never see a polar view. They all lie on their sides under the cover glass.

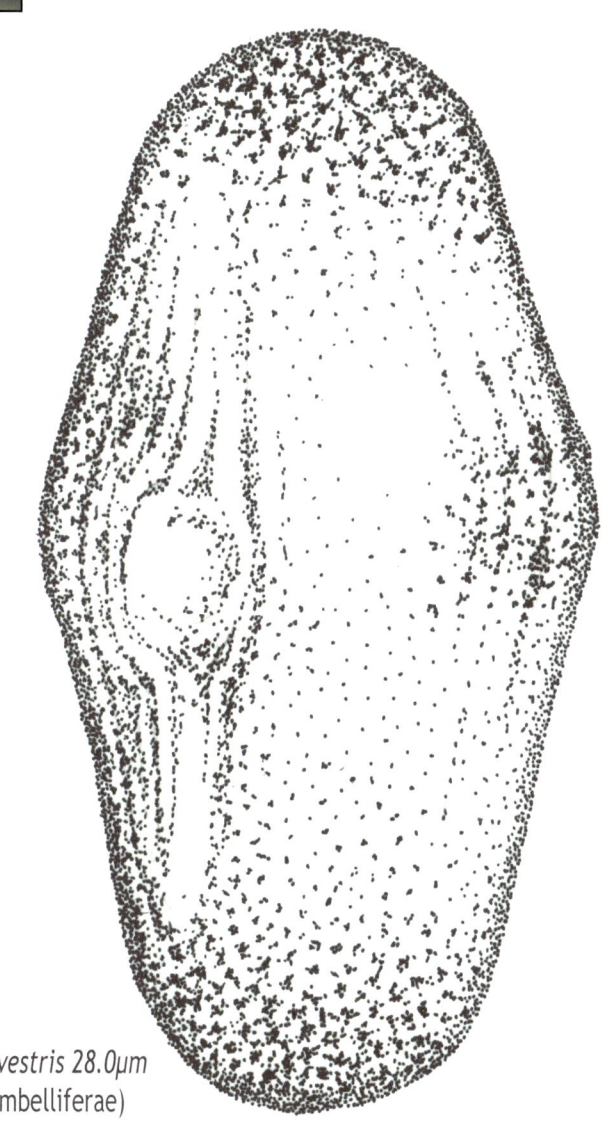

Anthriscus sylvestris 28.0μm
Apiaceae (Umbelliferae)

Cowslip

Cowslips grow readily in short grassland.
They hybridise readily with other close
members of the family.
The pollen grains have six or more
furrows, this grain has eight.

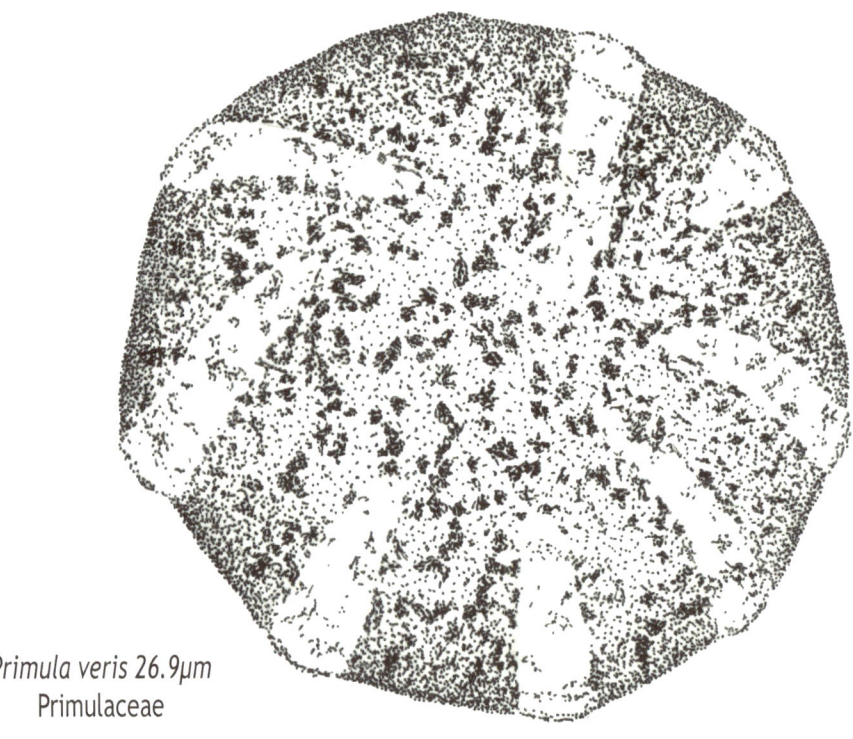

Primula veris 26.9μm
Primulaceae

Cranesbill

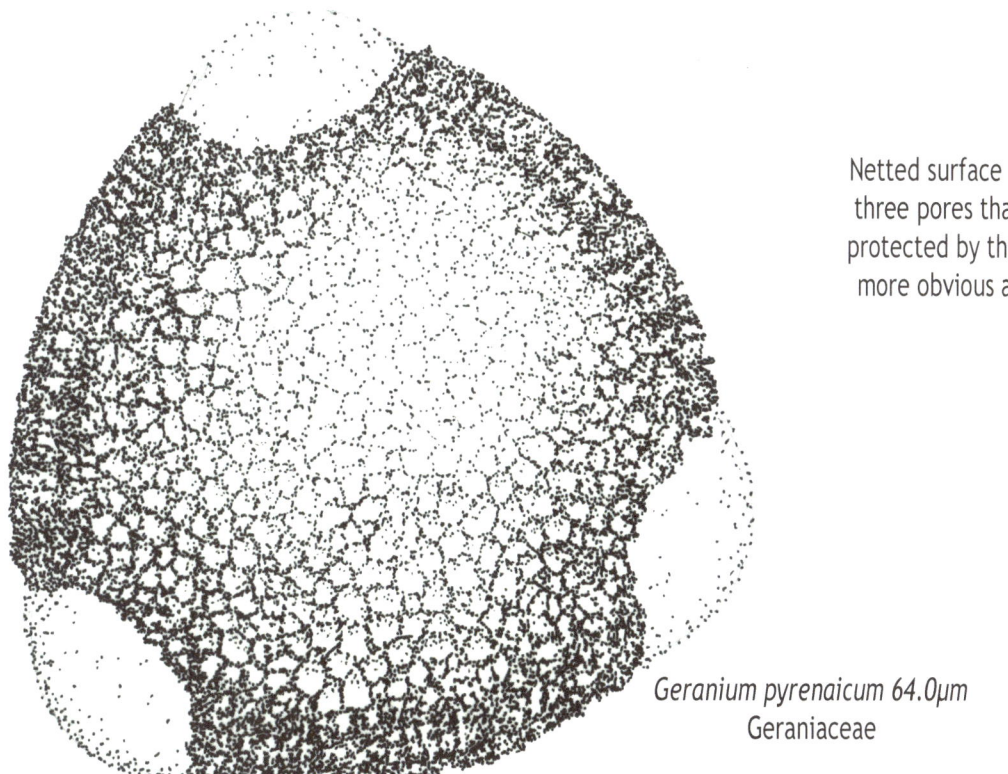

Netted surface of the pollen grain, with three pores that stay, within the grain, protected by the surface until becoming more obvious as the grain is hydrated.

Geranium pyrenaicum 64.0µm
Geraniaceae

Photo © Sally Dunn

Meadow Cranesbill

Large pollen grain with regular, reticulated surface and tube tips shown protruding through the three apertures.

Geranium pratense 101.0µm
Geraniaceae

Photo © Sally Dunn

Creeping bellflower

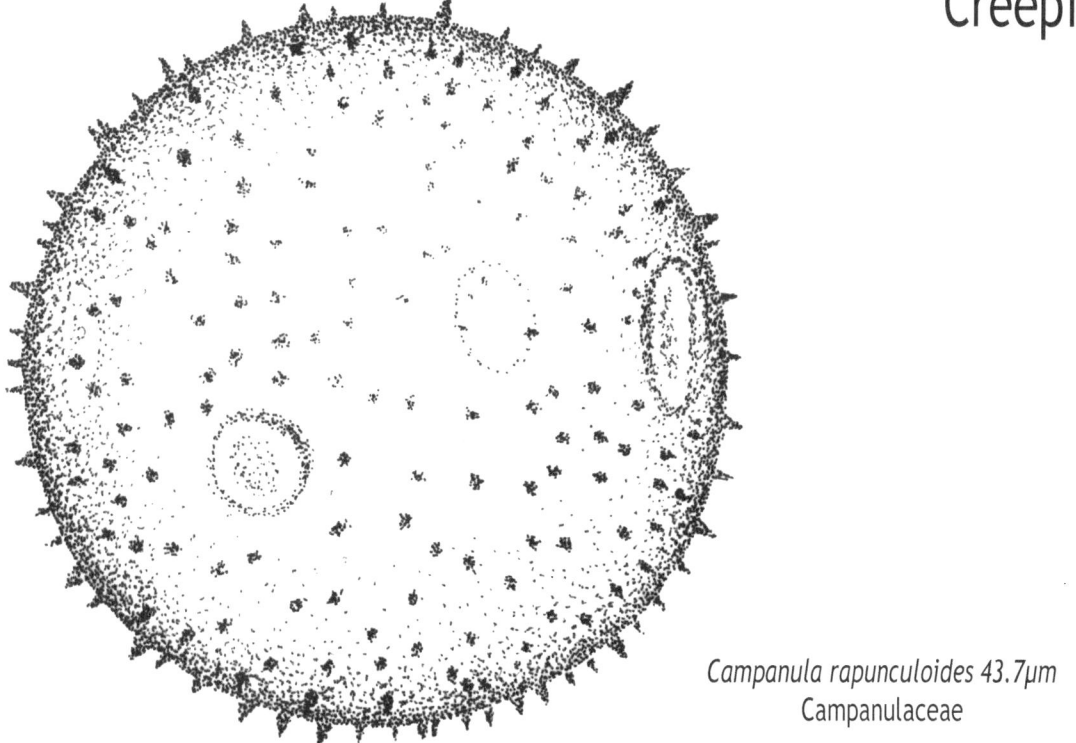

Campanula rapunculoides 43.7μm
Campanulaceae

The plant, though having attractive blue
flowers, is an invasive weed.
The pollen grain has up to five pores
at regular intervals around the equator
and is covered in tiny spines.

Creeping Jenny

The pollen grain has a regular but loosely patterned surface, and three furrows, each with a large pore at the equator.

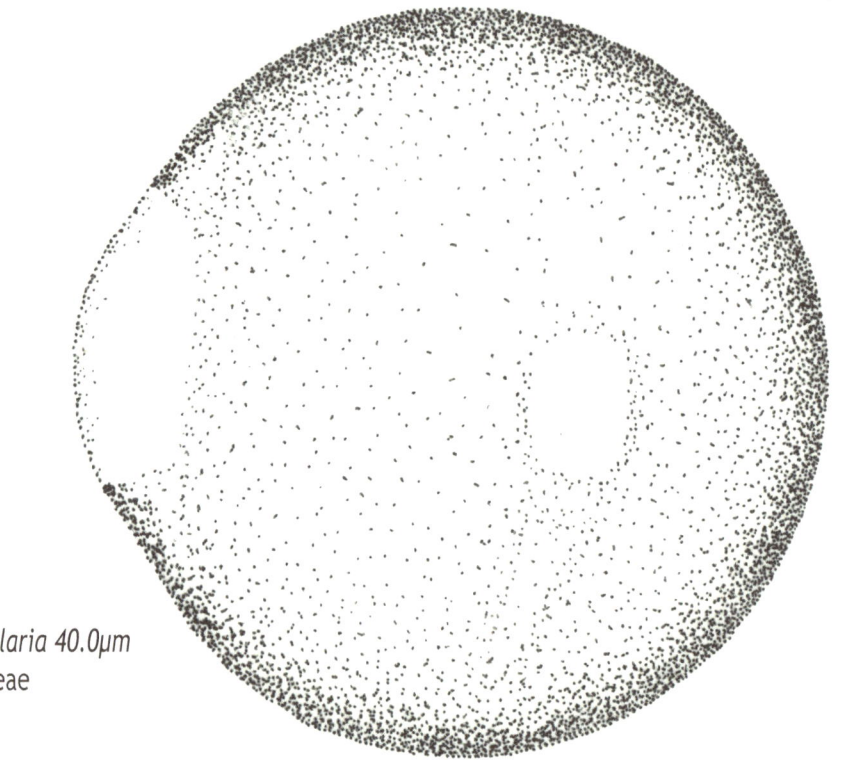

Lysimachia nummularia 40.0µm
Primulaceae

Photos © Sally Dunn

Spring crocus

Photo © Gerry Collins

Crocus sp. 115.0μm
Iridaceae

Spring crocuses are another good source of early pollen. They need a sunny spot as their flowers open only in sunlight, closing again at night. Their pollen grains are large and range in colour from yellow to deep orange.

Water crowfoot

A short-lived aquatic perennial with foliage submerged and white buttercup like flowers standing on or above the surface of the water. It is a good plant for pollinating bees, hoverflies and butterflies.
It has the typical pollen grain of the Ranunculaceae, spherical with three furrows.

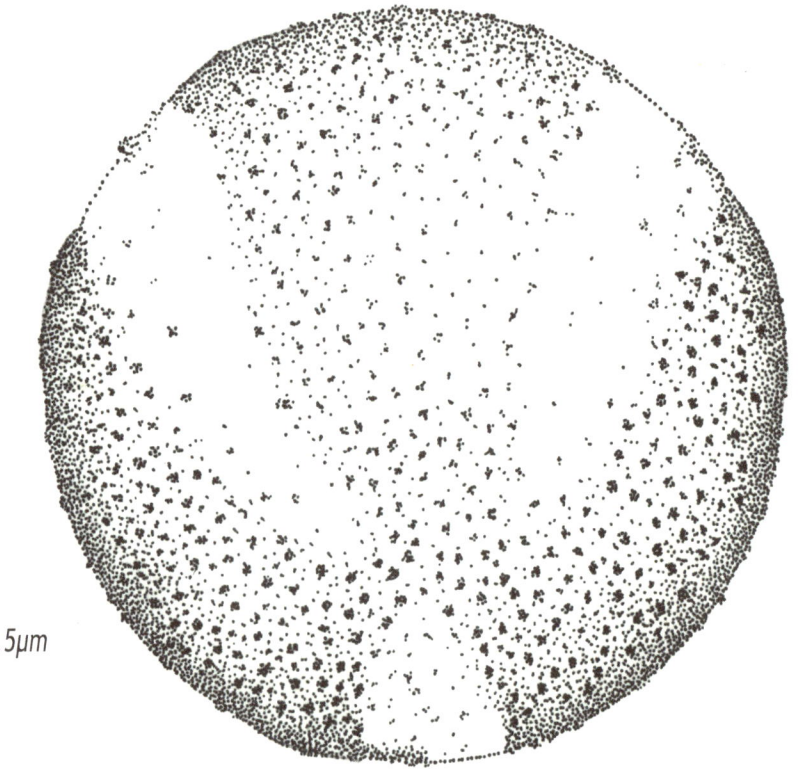

Ranunculus aquatilis 47.5µm
Ranunculaceae

Cyclamen hederifolium

Photo © Sue Carter

Cyclamen hederifolium (Cyclamen neapolitanum) 12.0μm
Primulaceae

Photo © Sally Dunn

A woodland plant, flowering in autumn, its pollen among the smaller sized grains.

Cypress

All conifers in the Cypress family produce extremely allergenic pollen.
Cypress pollen grains are spheroidal and any pore is not obvious.

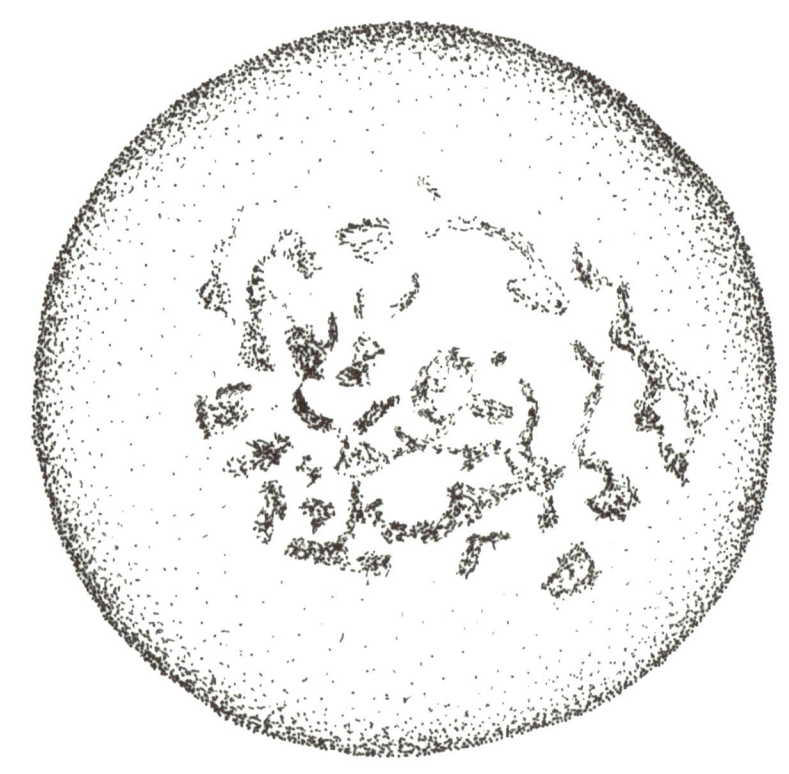

Cypress 29.8µm
Cupressaceae

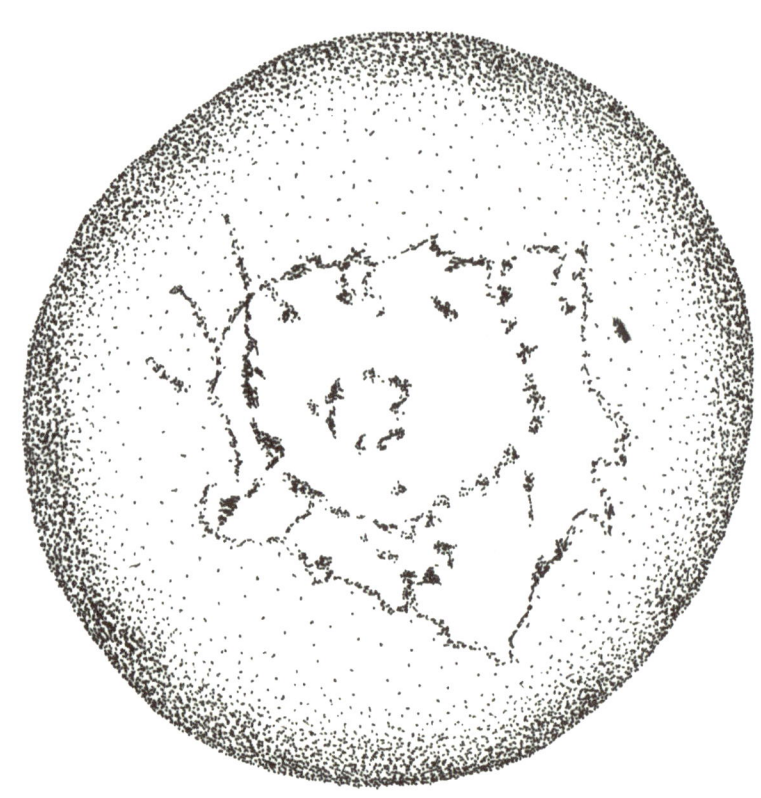

Daffodil

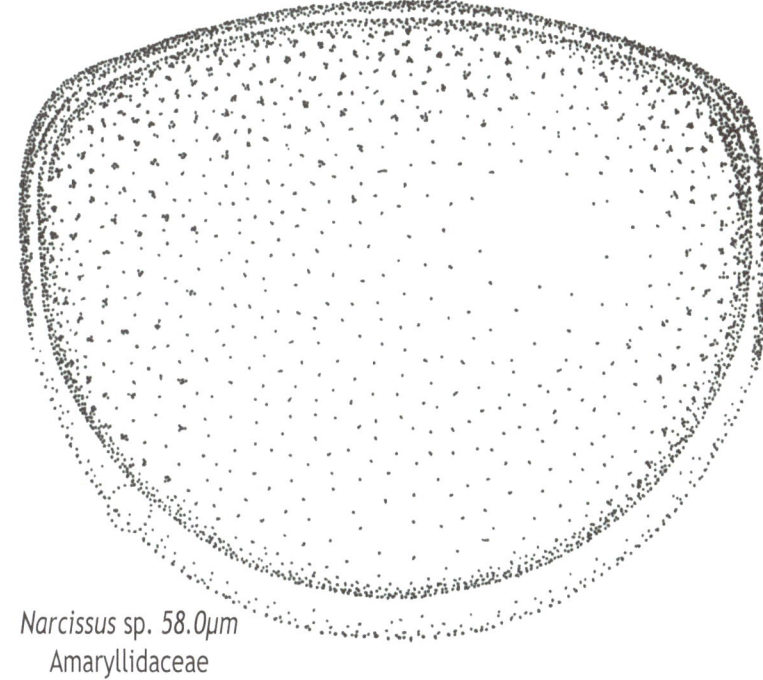

Narcissus sp. 58.0μm
Amaryllidaceae

The half moon shape of the pollen grain, equatorial view shown here, means one furrow, which is typical of the Amaryllidaceae.

Daisy

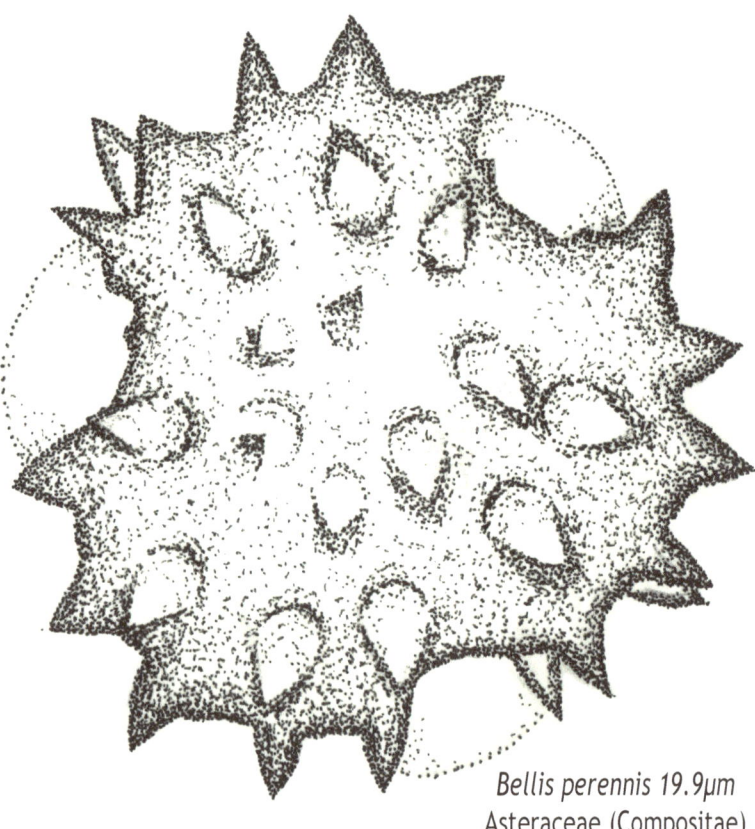

Bellis perennis 19.9μm
Asteraceae (Compositae)

The pollen grains have prominent broad based spines like nearly all the Compositae, and three pores.

Dandelion

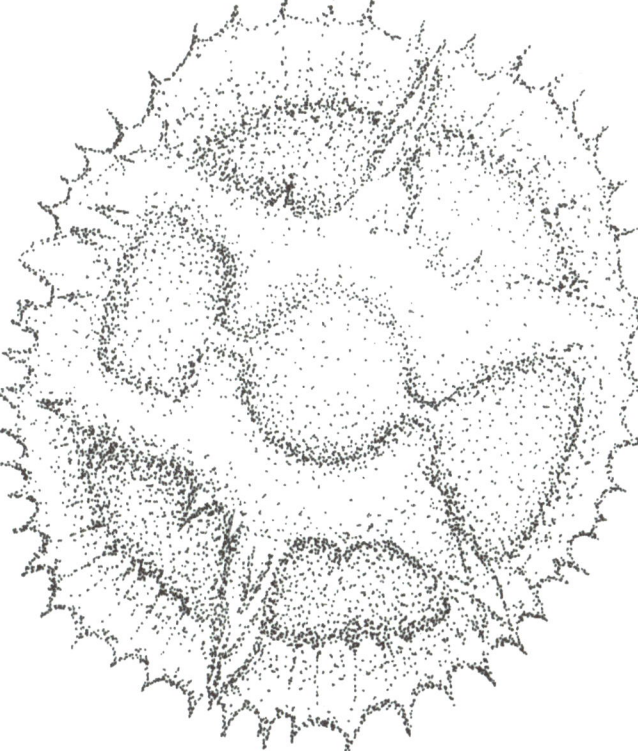

Dandelion pollen grains have an elaborate exine structure with ridges, valleys and three pores. Equatorial view above, polar view right. Note the bulging pores around the equator.

Taraxacum sp. 40.5µm
Asteraceae (Compositae)

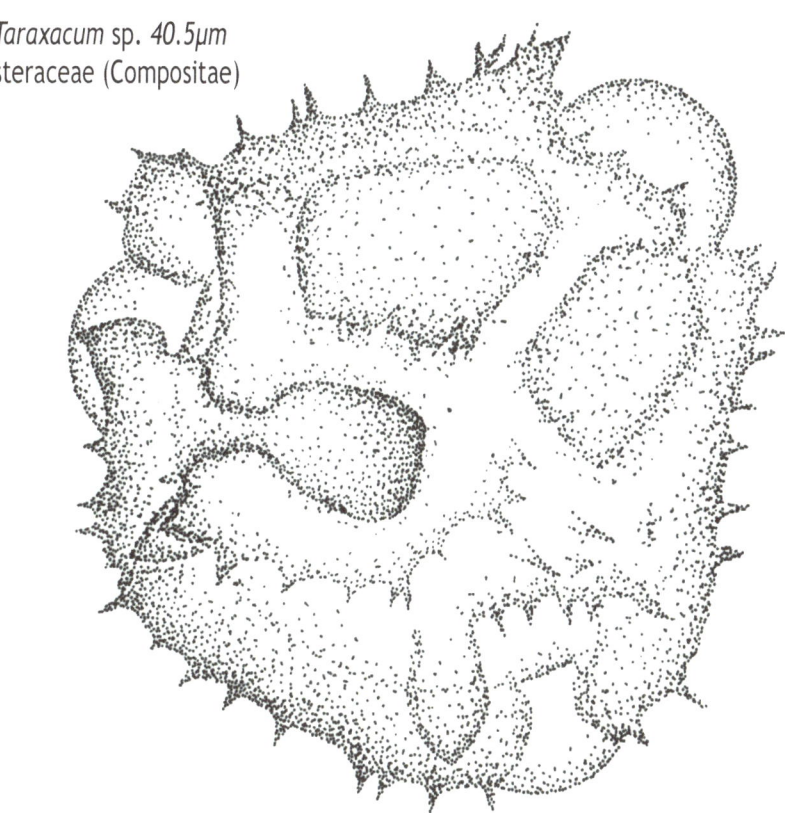

Dogwood

Photo © Sally Dunn

Equatorial view of the pollen grain. The three furrows are shallow and pores level with the surface, the top of a pore can just be seen top left.

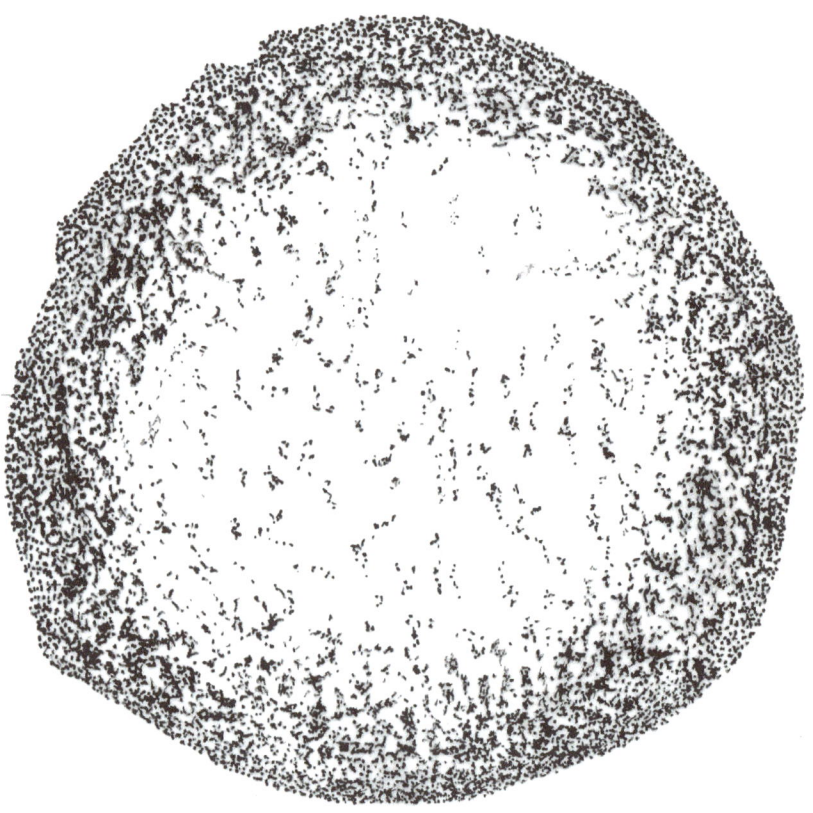

Cornus sanguinea 46.6μm
Cornaceae

Enchanter's Nightshade

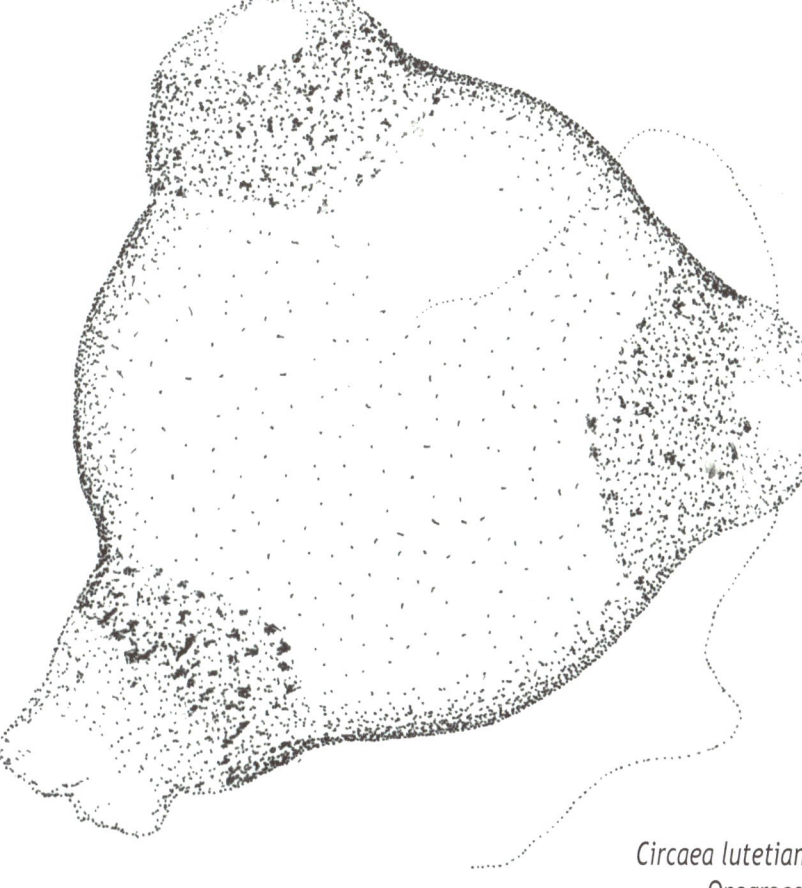

The pollen grain has quite a triangular shape with a prominent pore at each corner. Note the long vicin strand attached to the pole.

Circaea lutetiana 48.8μm
Onagraceae

Eucalyptus

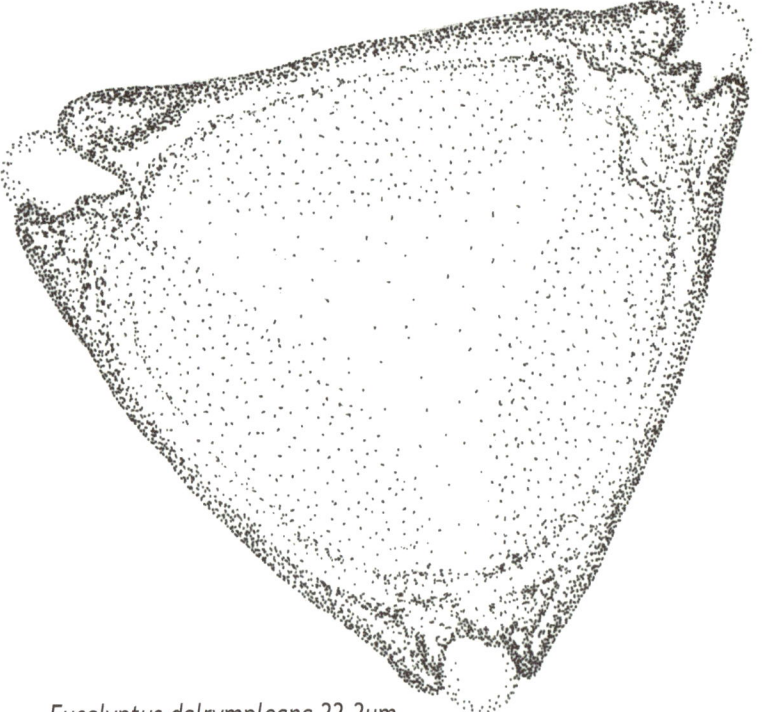

Eucalyptus dalrympleana 22.2µm
Myrtaceae

Each bud has a round cap composed of the fused sepals, petals or both. The flowers have no petals but as the white or red stamens expand and push through, they produce distinctive fluffy flowers consisting of the showy stamens. The flowers are very attractive to bees for both nectar and pollen.
The pollen grain is triangular with a small pore at each corner. The furrows join at the poles, forming a triangle.

Eucalyptus is a diverse genus of trees and shrubs with a distinctive flower formation.

Euphorbia (sun spurge)

Euphorbia helioscopia 32.4µm
Euphorbiaceae

An annual weed of cultivation. The three furrows almost meet at the poles. The surface of the grain is reticulated, as is the furrow, either side of the pore. The furrows often have smooth edges along their length.

Euphorbia (wood spurge)

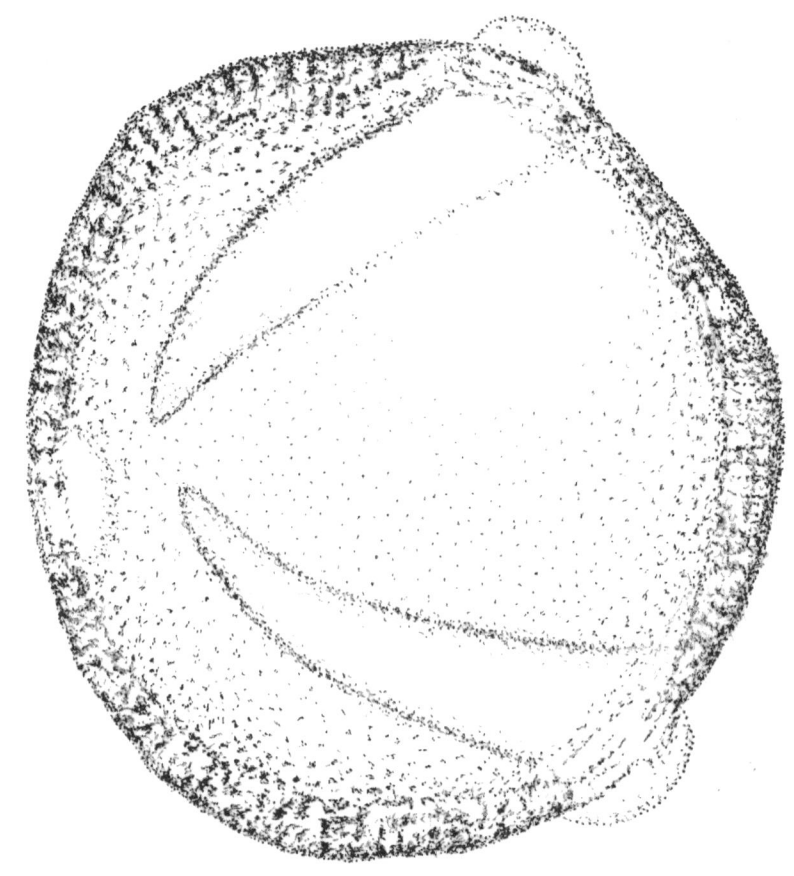

Euphorbia amygdaloides 40.0µm
Euphorbiaceae

An evergreen, woodland perennial,
growing to about 80cm with a compex
green yellow flower structure appearing
later spring early summer.
This angled polar view shows the lined
furrows and pores at the equator.

Evening primrose (large)

Oenothera erythrosepala 175.0μm
Onagraceae

With flowers opening late in the day, that is the time when pollen can be collected. The evening primrose is pollinated at night by insects such as moths, which fly later than is usual for honey bees. An efficient system has evolved using insects specially adapted for gathering these pollen grains. The grains are held together by sticky viscin threads, are relatively large in size, and low in quantity. The tangling together makes it almost impossible to find a separated pollen grain under the microscope.

Photo © Sally Dunn

Fatsia

Rather spectacular evergreen shrub with distinctive, waxy, lobed leaves. Compact flowerheads are borne in autumn followed by round black fruit. The pollen grain has a reticulated surface, and short furrows with pores that are seen when the grain is hydrated.

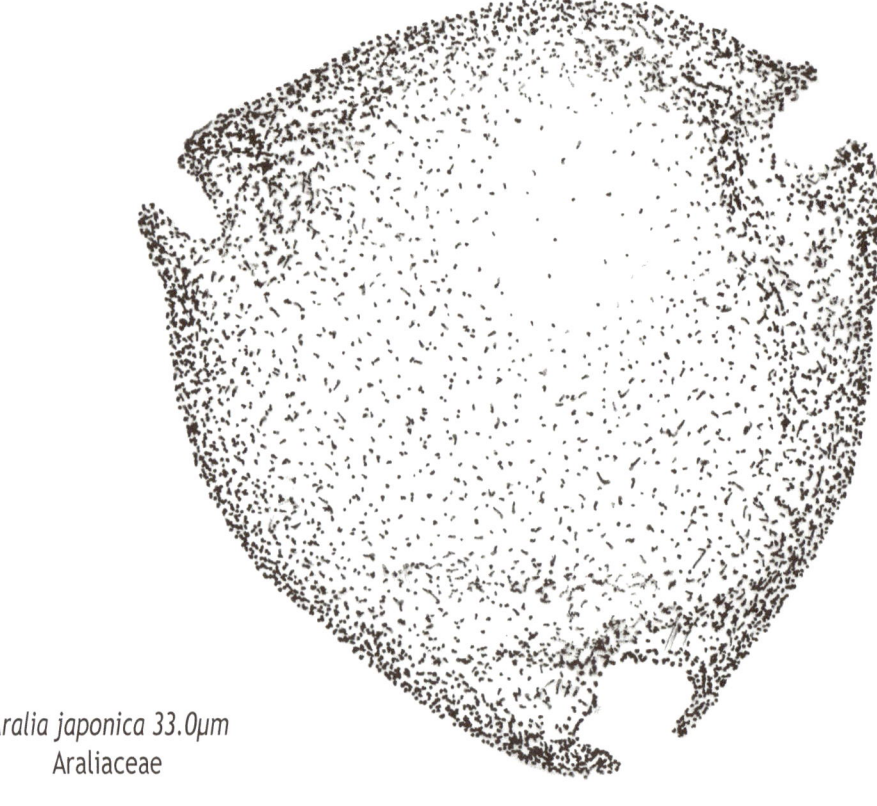

Aralia japonica 33.0µm
Araliaceae

Field Mouse-ear

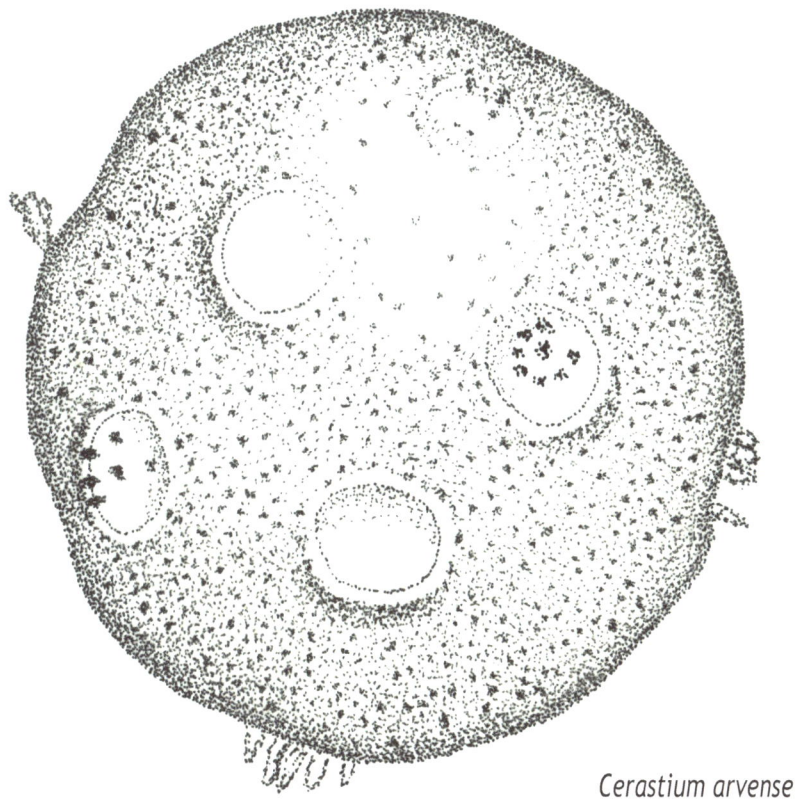

Pollen grain with tiny pores and dots sticking upwards all over the surface. Several, up to 20, large pore apertures are evenly spaced over the sphere, and themselves have a decorated surface.

Cerastium arvense 33.5µm
Caryophyllaceae

Photo © Sally Dunn

Forget me not

Photo © Gerry Collins

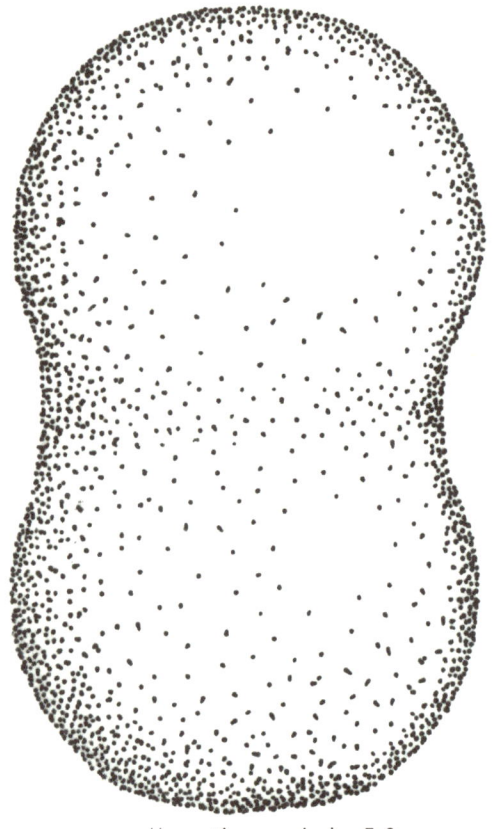

Myosotis scorpiodes 5.8µm
Boraginaceae

One of the smallest pollen grains, usually around 6µm, with its characteristic dumbbell shape. The three furrows, seen under an electron microscope are not visible on such a tiny grain under a light microscope.

Forsythia

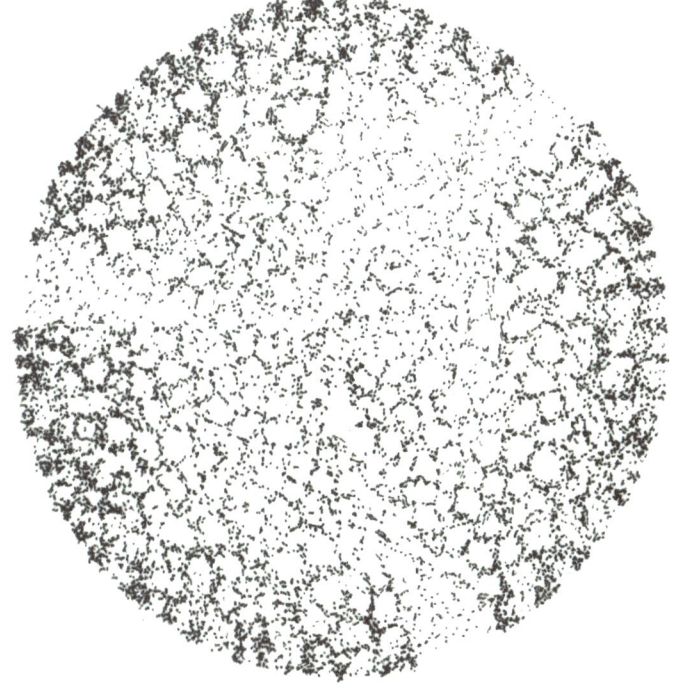

Forsythia intermedia 23.0µm
Oleaceae

Shrub, sometimes used as a hedge, covered with yellow flowers in spring. Polar view of the netted pollen grain above, equatorial view right. A smallish grain which is quite spherical enabling both polar and equatorial viewing from the microscope slide.

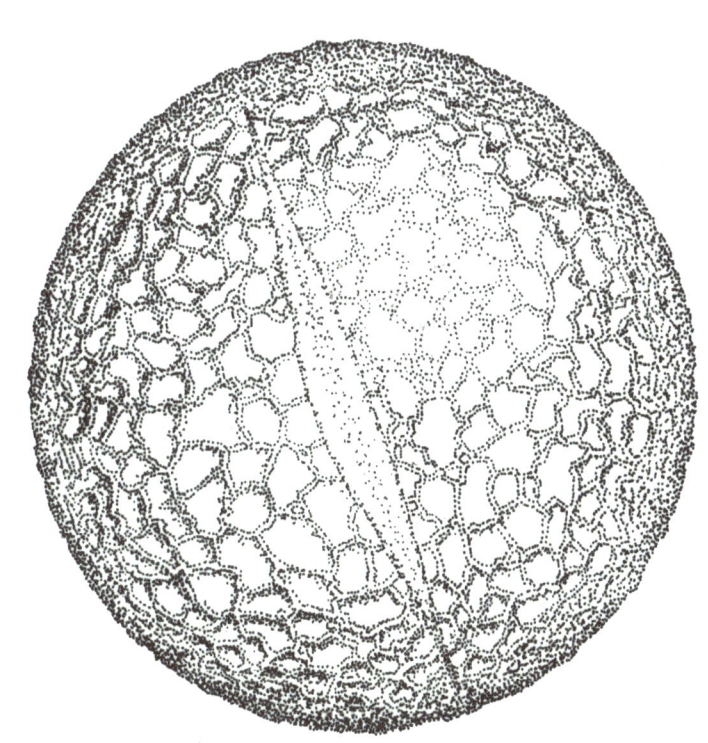

Foxglove

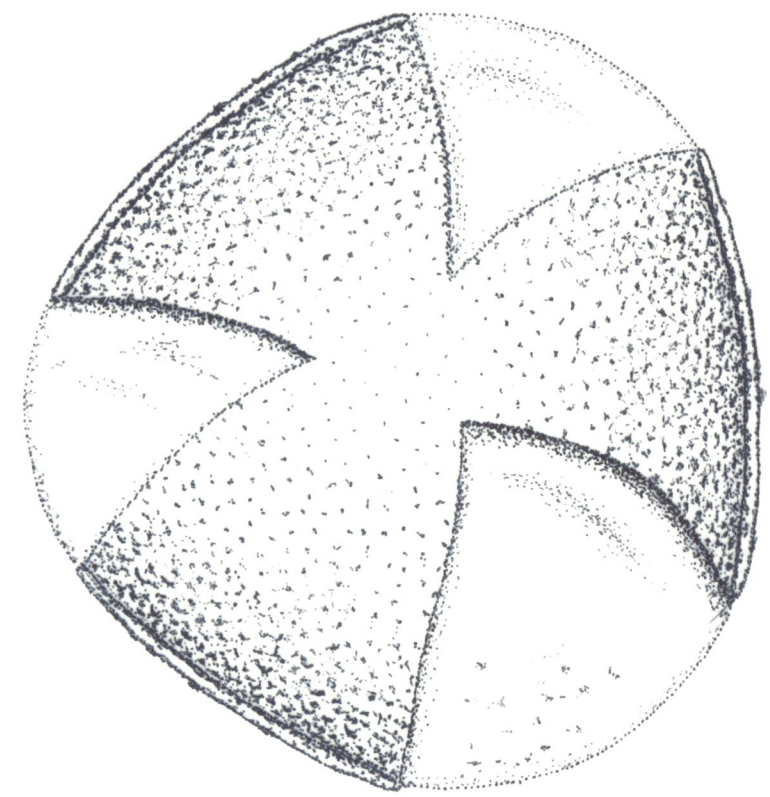

Digitalis purpurea 25.0μm
Plantaginaceae

Pollinated mainly by bumble bees, which have to crawl up the long tube, brushing past the stamens, to get to the nectar at the base.

Photo © Sally Dunn

Freesia

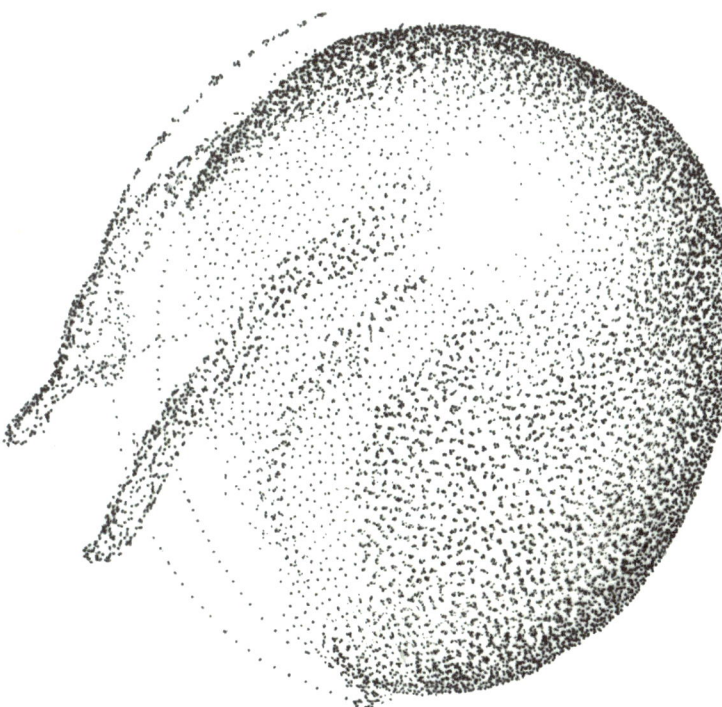

A frost tender perennial grown for its fragrant flowers.
Polar view of the pollen gain.
The pollen grain is quite large, and elongated with one furrow.

Freesia (cut flower) 84.6µm
Iridaceae

Photo © Sally Dunn

Gladiolus

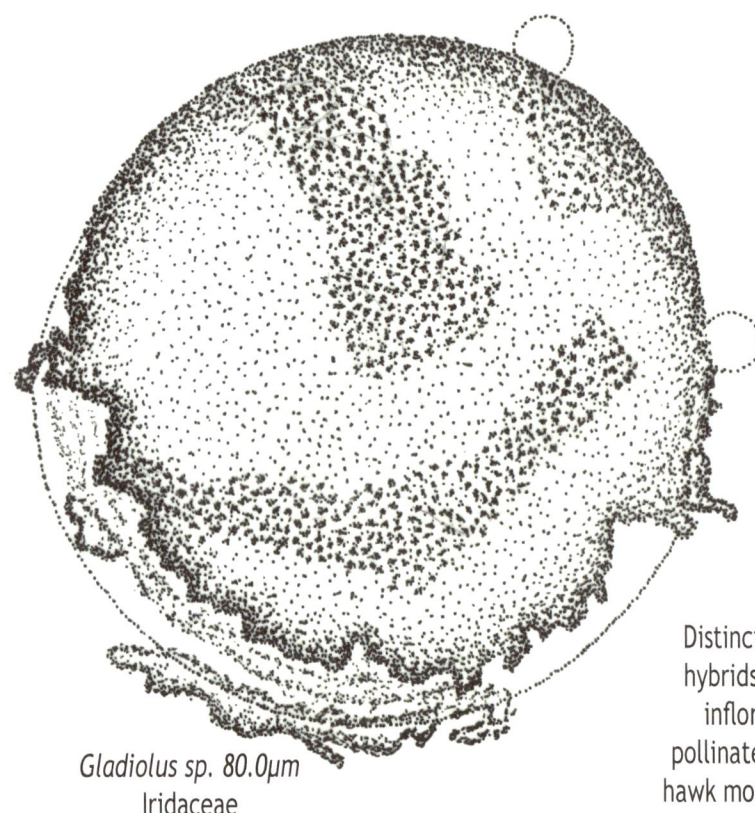

Gladiolus sp. 80.0μm
Iridaceae

Distinctive garden hybrids with large inflorescences pollinated mainly by hawk moths and small wasps.
The borders of the single furrow are highly decorated taking up almost a third of the surface of the grain. The thin exine is breaking up in this illustration.

Canadian golden rod

Photo © Gerry Collins

Solidago canadensis 22.8μm
Asteraceae (Compositae)

Like most in the daisy family, the exine of the pollen grain bears several broad based spines.

Photos © Sally Dunn

Red goosefoot

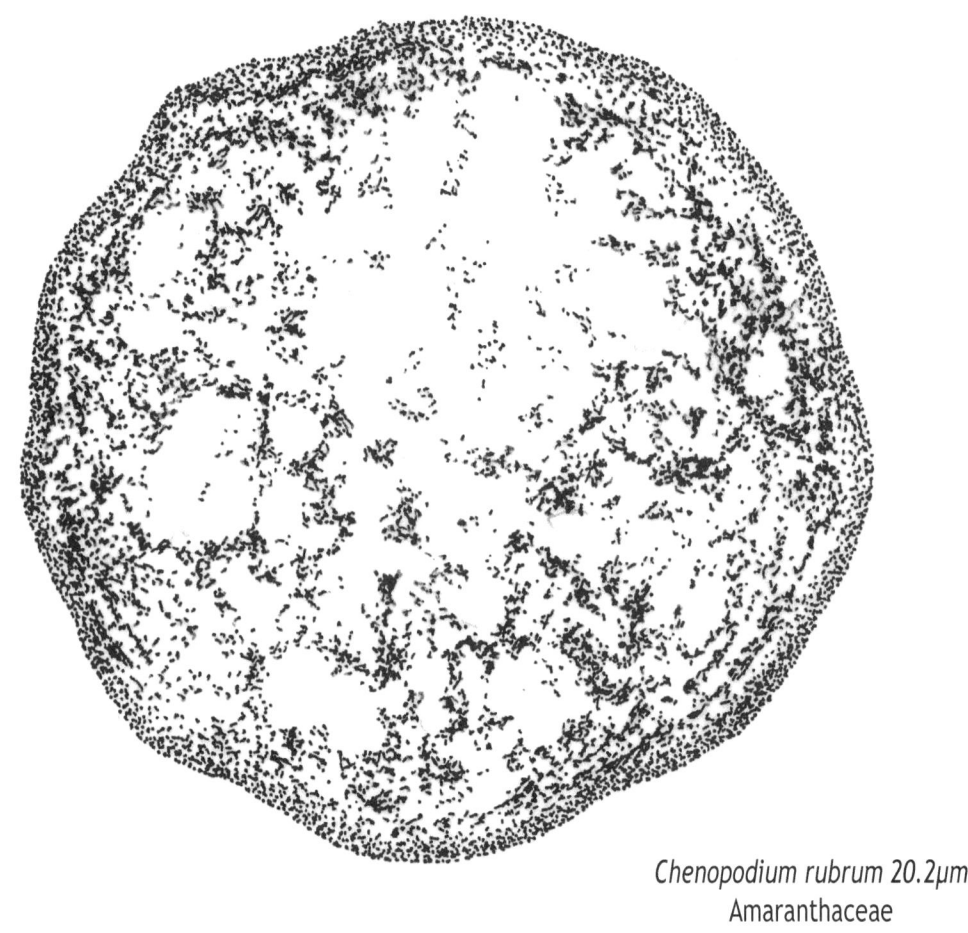

Chenopodium rubrum 20.2μm
Amaranthaceae

An annual weed, a member of the
quinoa family. It likes nitrogen rich soil.
The pollen grain has pores at regular
intervals over the whole surface.

Gorse

Ulex europaeus 40.8µm
Fabaceae

A prickly, shrubby bush often seen colonising sandy soil. It is insect pollinated, but not one of the easiest of flowers for teasing out the pollen for microscopy.

Grape hyacinth

Muscari armeniacum 35.3µm
Asparagaceae

Photo © Sue Carter

Extracting the pollen from these tightly packed grape hyacinth flowers presents a challenge for the microscopist, but not, it seems, for the bees. The pollen grain has one furrow which almost meets up with itself.

Greater spearwort

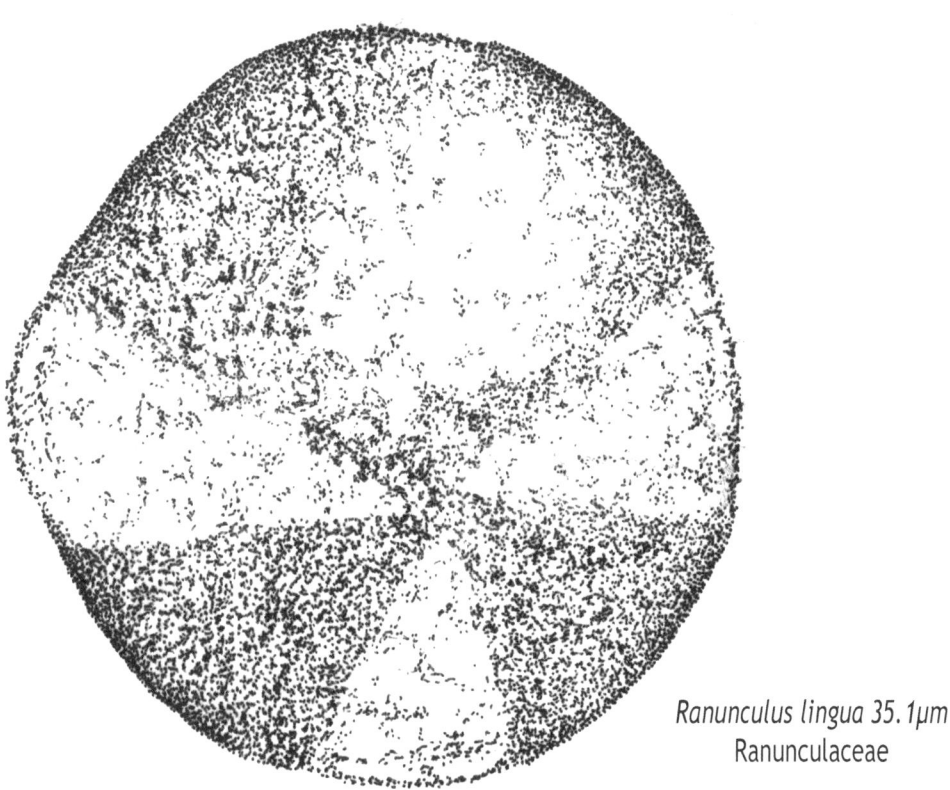

Ranunculus lingua 35.1μm
Ranunculaceae

A marginal aquatic plant in the buttercup family with a flower very similar to the common buttercup. This illustration shows three furrows meeting. The grain usually has between six and twelve furrows.

Groundsel

A persistent weed of cultivation. Its pollen grains have broad based spines. The illustration, right, shows the larger part of a furrow. It will be found wide open or closed depending upon the amount of water absorbed, which can swell or shrink a grain. This grain exhibits the spiky protrusions characteristic of the daisy family.

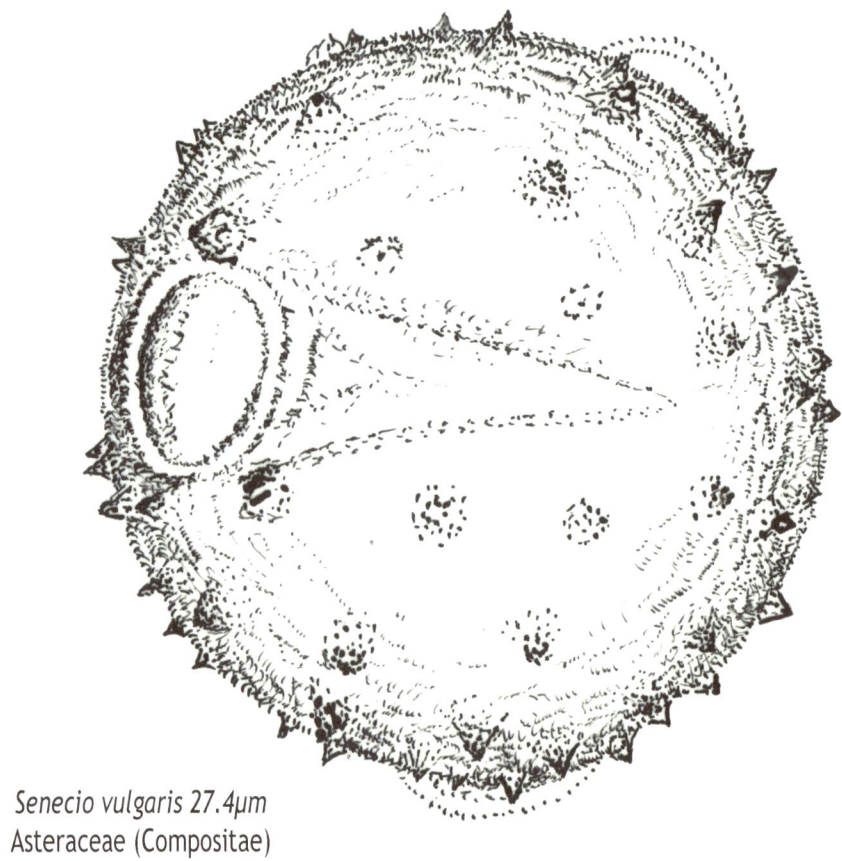

Senecio vulgaris 27.4μm
Asteraceae (Compositae)

Harebell

Photo © Sally Dunn

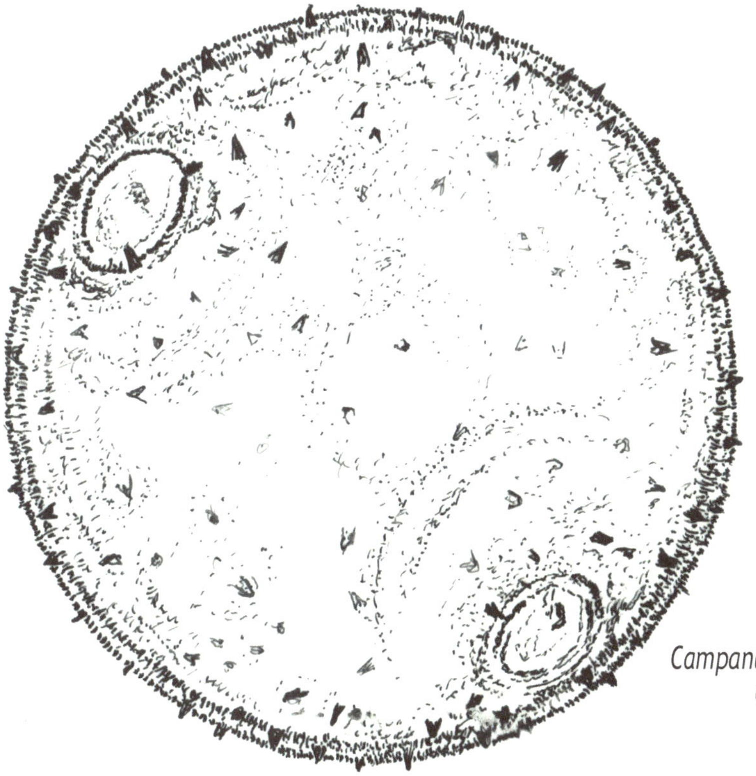

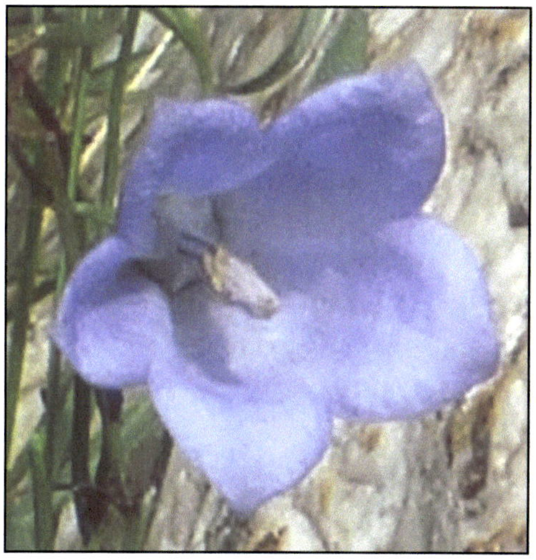

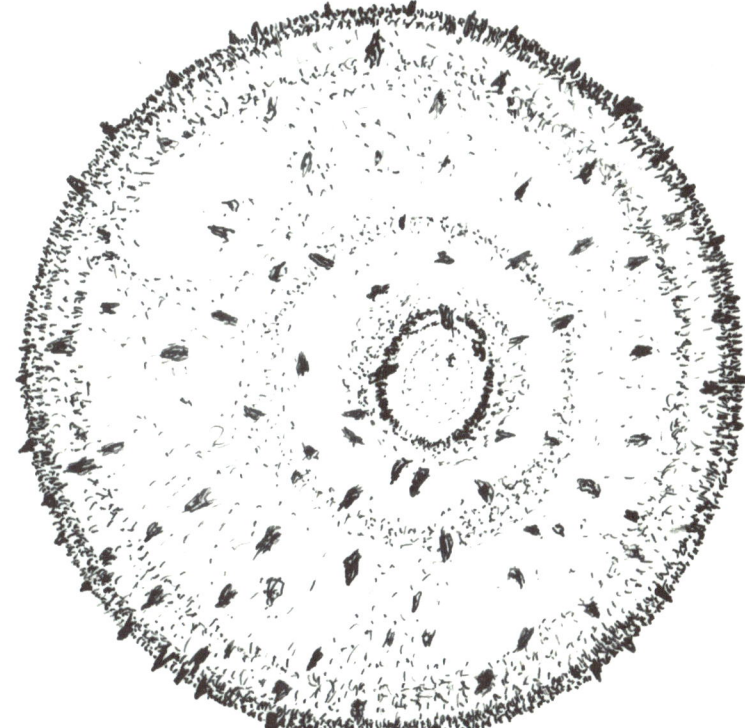

Campanula rotundifolia 40.0μm
Campanulaceae

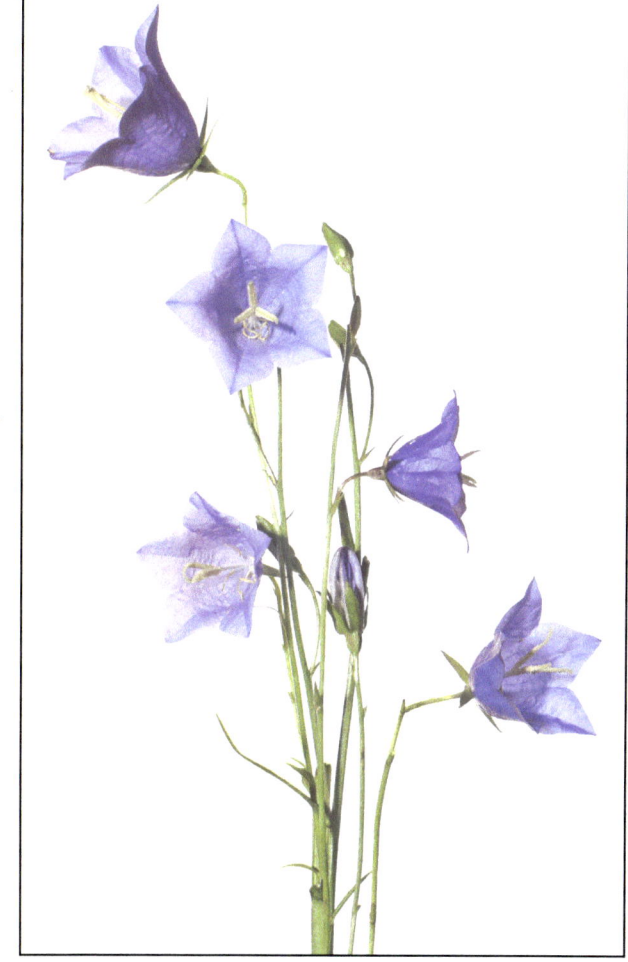

Harebells have transparent pollen grains. The top view pollen drawing shows two pores, the view below is the same grain, the rear face viewed from the inside showing the third pore. Not all species have transparent grains.

Hawthorn

Thorny, hardy, large shrub, often used for hedging. Profuse flowers are produced in May. The freely exposed stamens provide food for a very wide variety of insects from midges to butterflies.

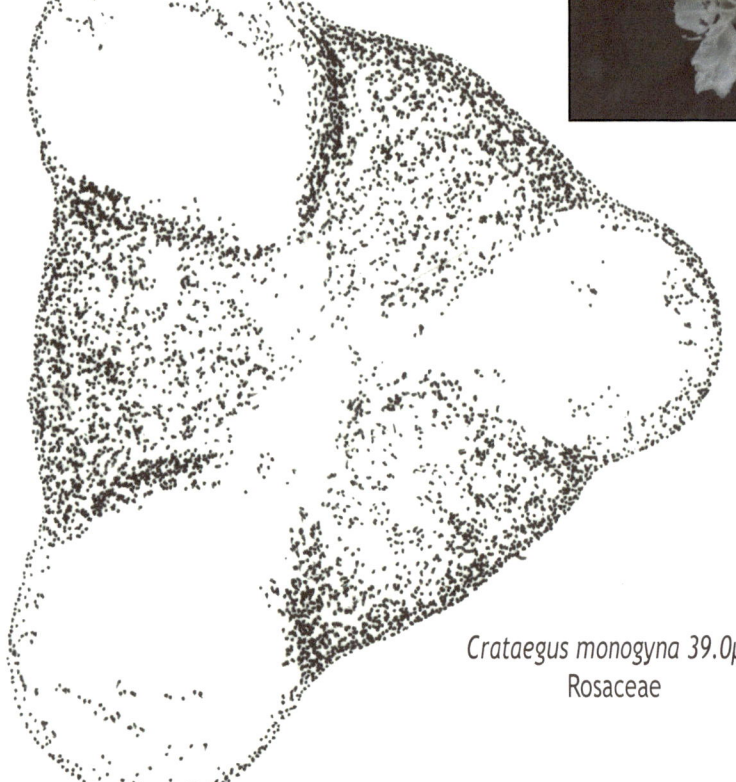

Crataegus monogyna 39.0μm
Rosaceae

Hazel

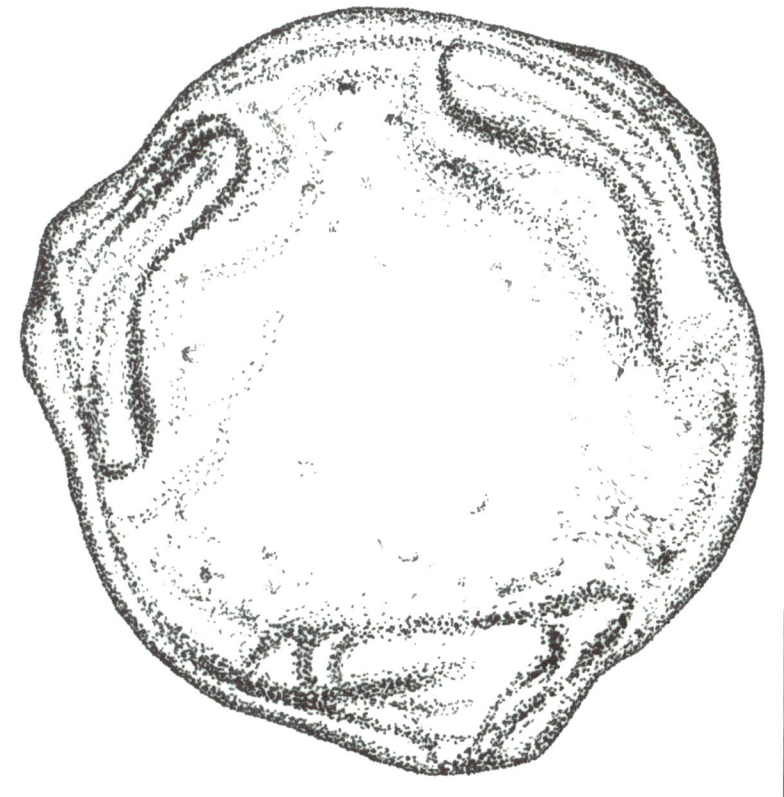

Corylus avellana 25.4μm
Betulaceae

Hazel trees' pollen-bearing male catkins are long and quite noticeable. Female flowers are tiny, red and not so obvious. Hazel provides a good, early source of pollen, though for the microscopist, whilst easy to collect, the grains tend to bunch under the cover glass.
The size of the hazel pollen grain is fairly constant and used as a reference for pollen grain sizes. Its shape is oblate so lies flat with the view under the microscope always a polar view.

Bell heather

Erica cinerea 64.4µm
Ericaceae

Photo © Sally Dunn

Heather flowers late in the season, August and September, and is very attractive to bees. The honey has a unique taste and texture. It is thixotropic, almost solid when undisturbed, becoming temporarily liquid when stirred or agitated.
The pollen grains aggregate in clusters of four.

Cross leaved heather

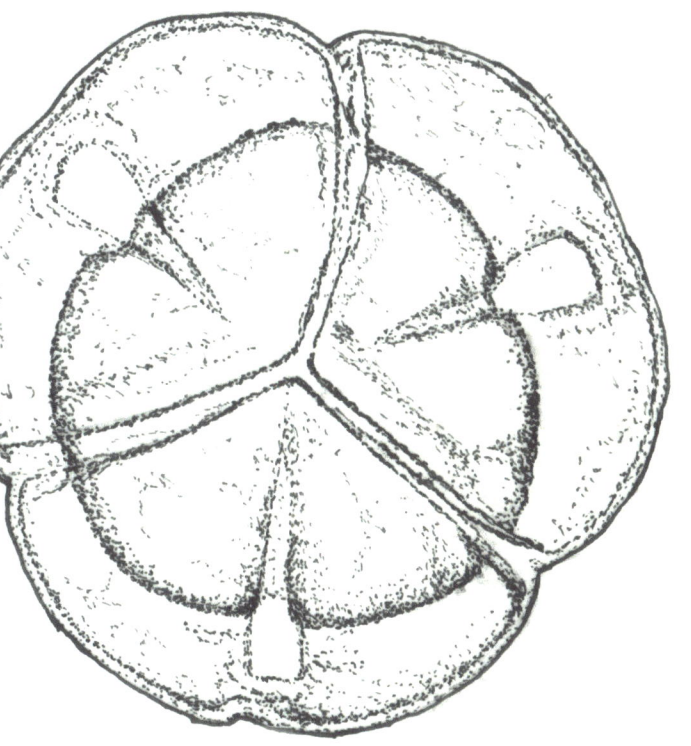

Erica tetralix 45.8μm
Ericaceae

The leaves of cross leaved heather are arranged on the stem in whorls of four, and pink bell shaped flowers are in a small cluster at the tip of long stems. This illustration shows the aligned furrows in the polyad cluster typical of the family.

Ling heather

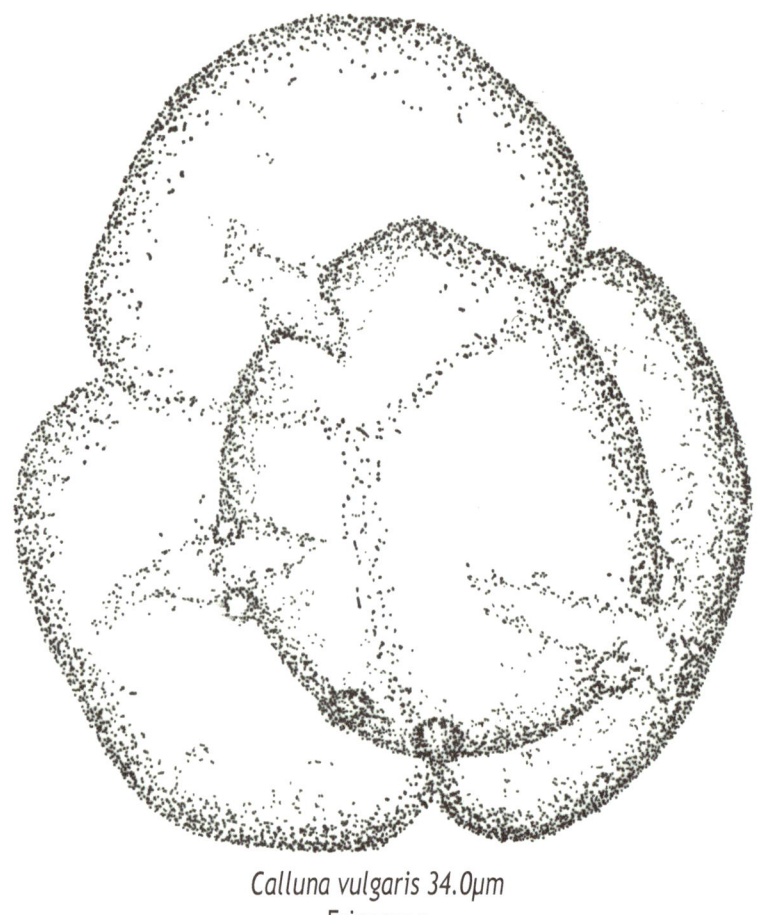

Calluna vulgaris 34.0µm
Ericaceae

Ling, or common heather, often just called 'heather' is a low growing shrubby bush which grows on open, acid soil. The leaves, in opposite pairs, are smaller than those of other heathers.
The tetrad pollen formation is typical of the the Ericaceae.

Hedge mustard

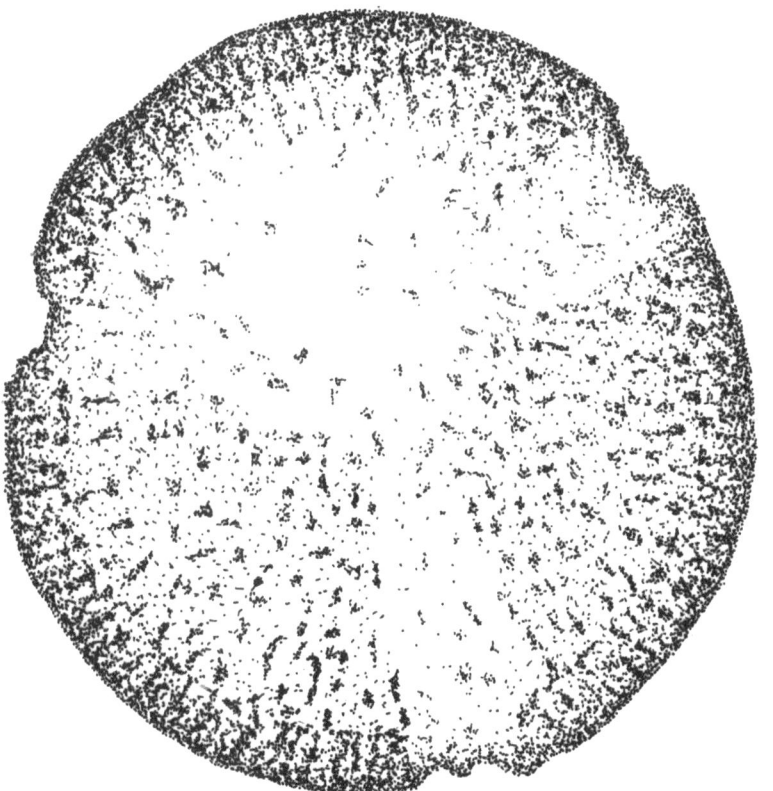

Sisymbrium officinale 25.4μm
Brassicaceae (Cruciferae)

The pollen grain has a neatly latticed surface and three furrows which have even width along their length, not quite meeting at the poles.

Hellebore

Strange flowers that have a ring of small green tubular 'honey leaves' around the stamens supplying a generous supply of nectar. Flowering in winter, hellebores are a good source of early pollen, much needed by honey bees to feed their brood ready for the spring and summer. The pollen grain has three wide furrows, with the furrows more noticably decorated than the rest of the pollen grain surface.

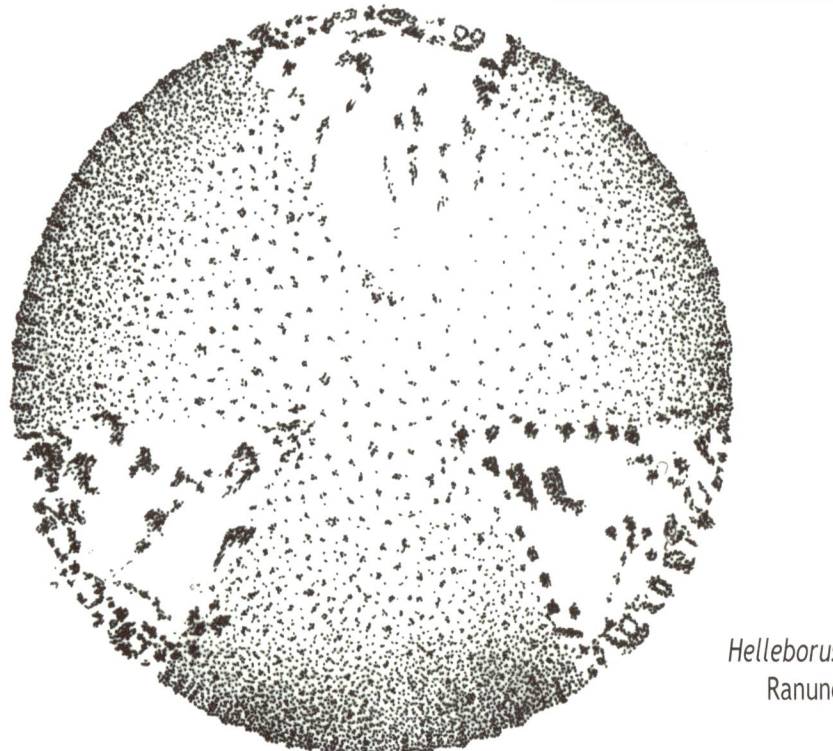

Helleborus sp. 34.5µm
Ranunculaceae

Herb Robert

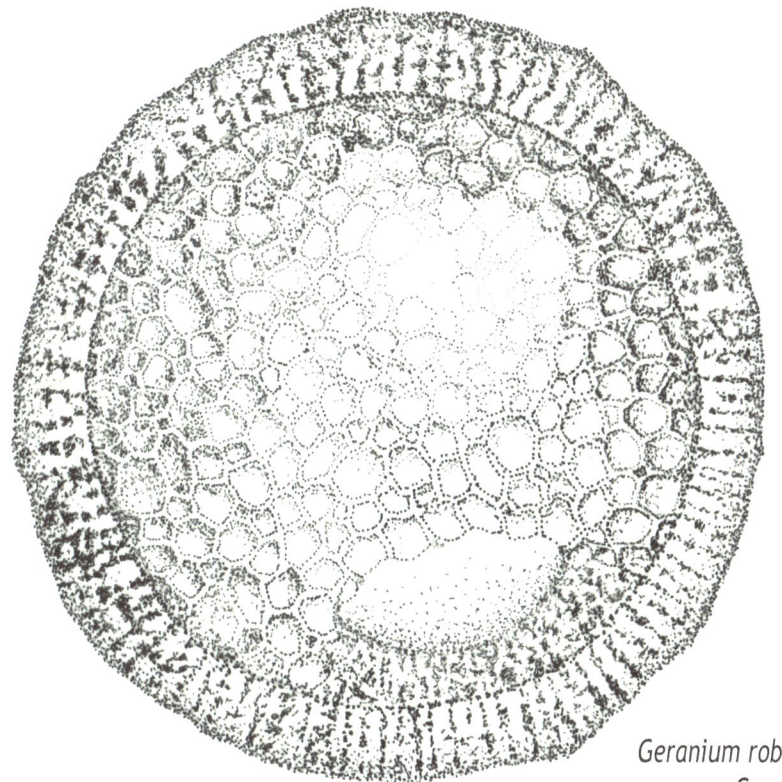

A ubiquitous weed, the pollen grain has three pores and is net patterned.

Geranium robertianum 71.1µm
Geraniaceae

Photo © Sally Dunn

Hesperis (dame's violet, sweet rocket)

Hesperis matronalis 22.2µm
Brassicaceae (Cruciferae)

Tall straggly short lived perennial of the brassica family, which self seeds readily. It has many common names including 'dame's violet' and 'sweet rocket'. The flowers range between white and violet and are heavily scented. The pollen is netted with visible furrows. It bunches under the cover glass.

Hibiscus

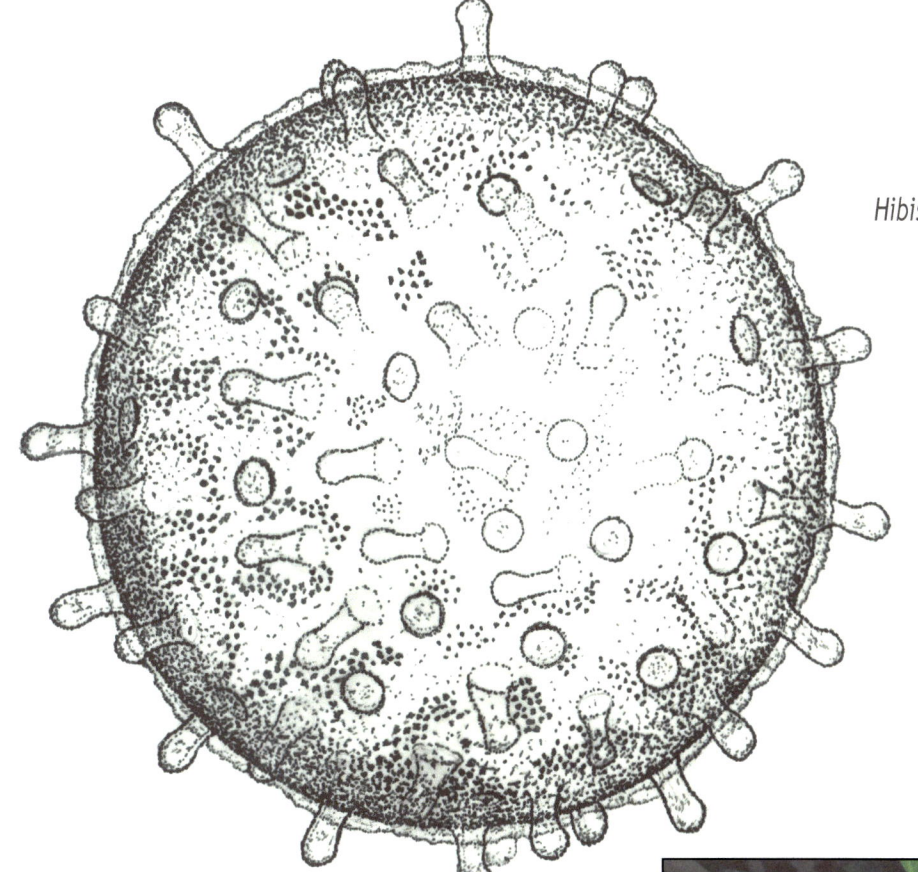

Hibiscus 'The President' 200.0µm
Malvaceae

One of the largest pollen grains, spherical with spiny, odd shaped protrusions interspersed with pores over the entire surface. Similar features can be seen in the pollen of other members of the mallow family.

Hoary cress

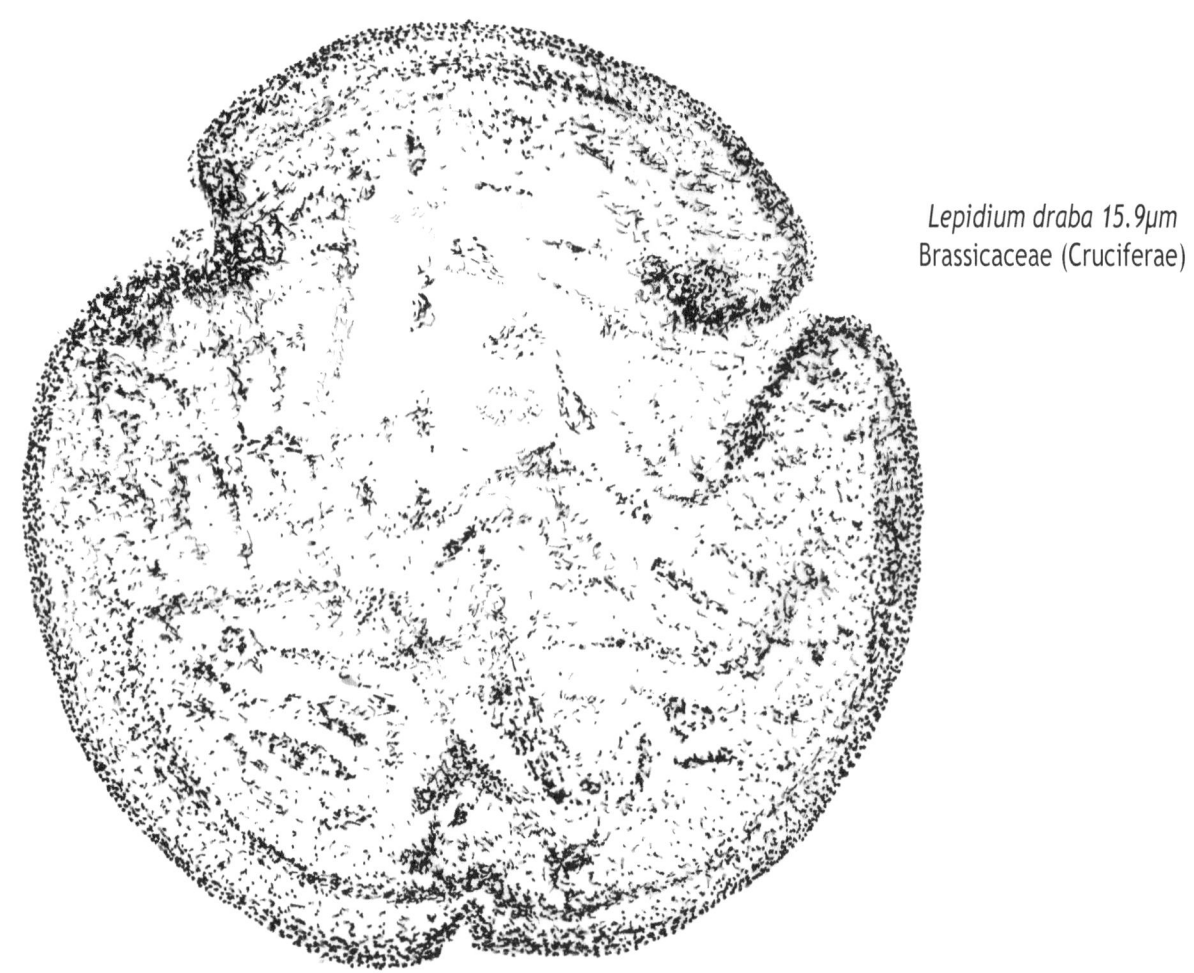

Lepidium draba 15.9µm
Brassicaceae (Cruciferae)

A perennial plant with a spread of white flowers across the top of the plant, giving it a common name of whitetop. The grain has three furrows. This polar view shows sunken furrows implying the grain is dry.

Hogweed

Photo © Janet Morris

Tall, perennial wild flower with a characteristic smell, pollinated by insects including wasps and flies. The three slit like furrows don't reach the poles. Each furrow has a pore at the centre.

Heracleum sphondylium 40.5µm
Apiaceae (Umbelliferae)

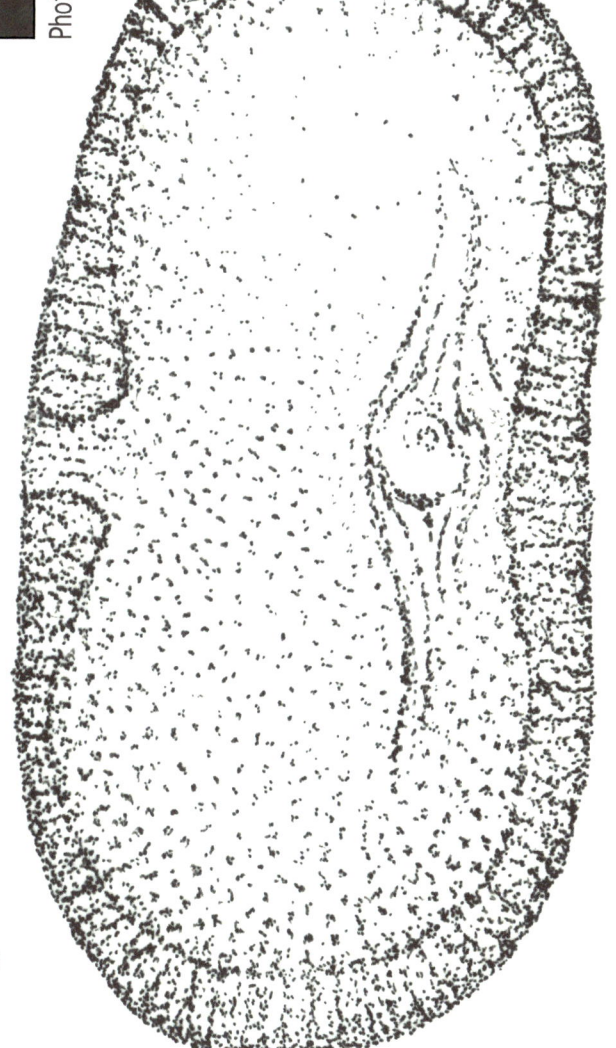

Holly

Male and female flowers are borne on separate trees. The male flowers produce pollen, and only the female trees produce berries, if there is a pollinator tree nearby, and if there are insects to transport the pollen. The exine of the pollen grain shows a unique knobbly surface. Compare this with that of red valerian which houses a similar effect, but on the furrows.

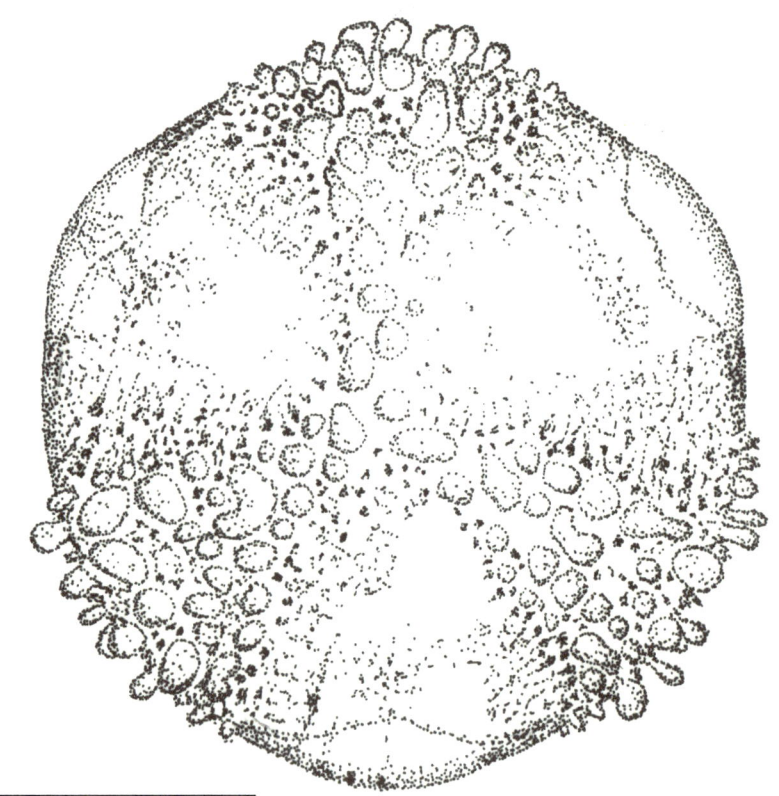

Ilex aquifolium 44.7μm
Aquifoliaceae

Hollyhock

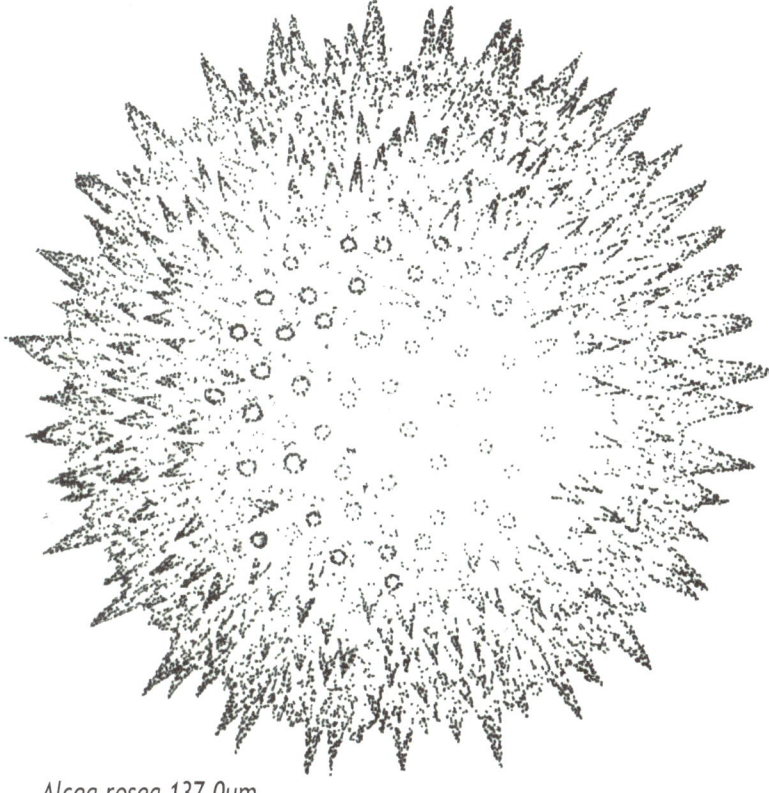

Alcea rosea 137.0μm
Malvaceae

Tall stems of large flowers shedding copious quantities of pollen. Visiting insects can be seen sometimes conpletely dusted with the pollen.
The pollen grains are among the largest in size, covered in spines, interspersed with numerous pores.

Honesty

Lunaria rediviva 21.6µm
Brassicaceae (Cruciferae)

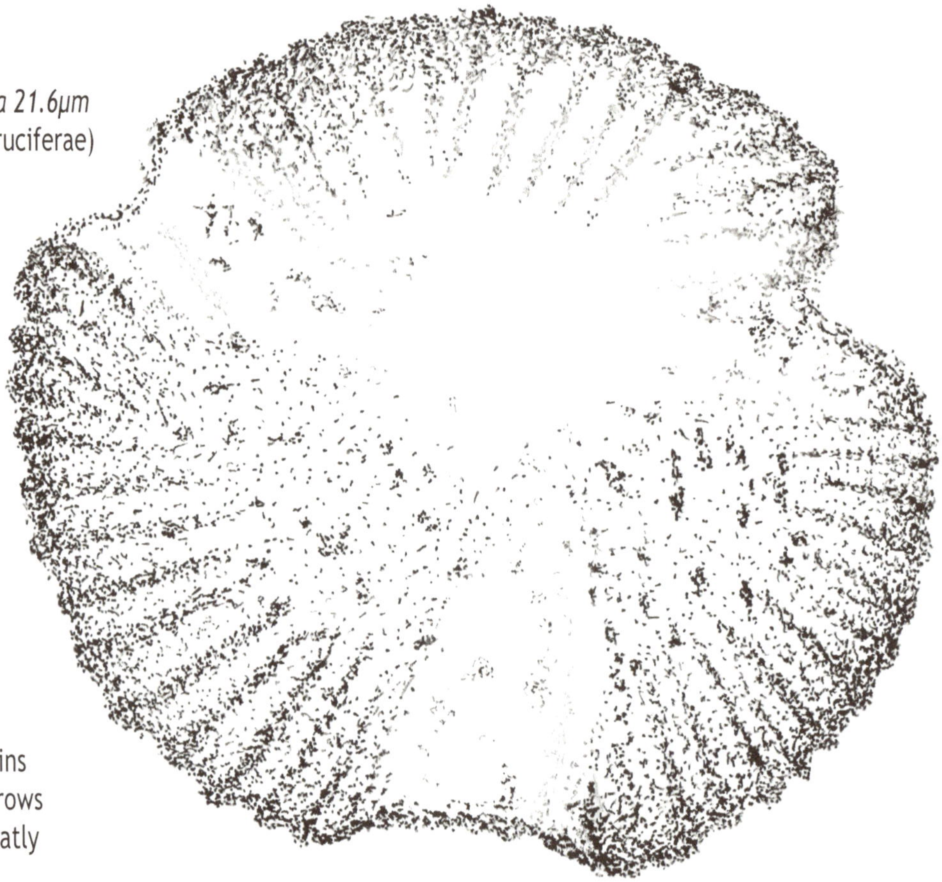

Perennial honesty pollen grains have three slightly sunken furrows and a loose lattice pattern neatly aligned as the lines show.

Honeysuckle

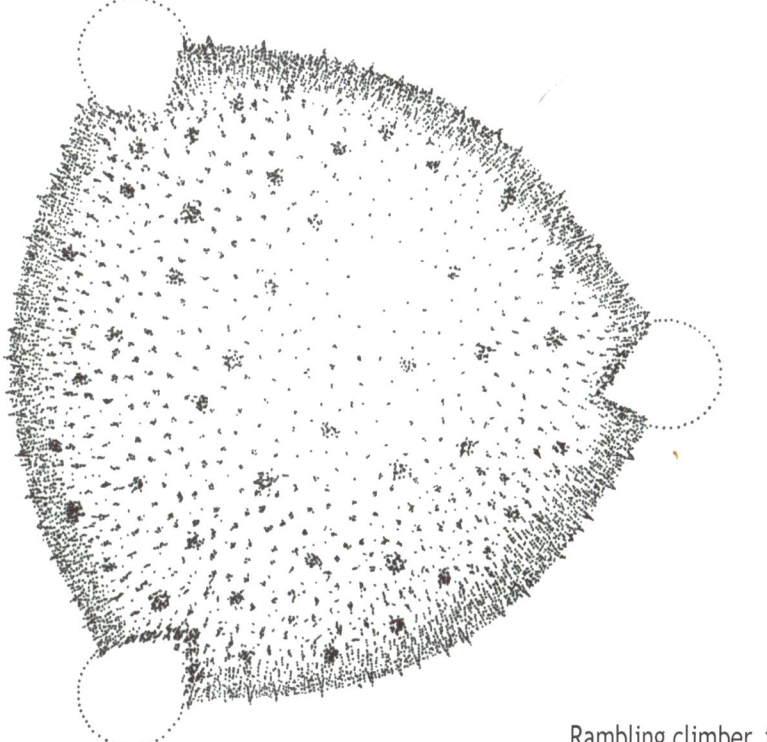

Lonicera periclymenum 62.0μm
Caprifoliaceae

Rambling climber, the anthers are prominent, making pollen collection easy, it can be smeared straight on to the microslide.

Hornbeam

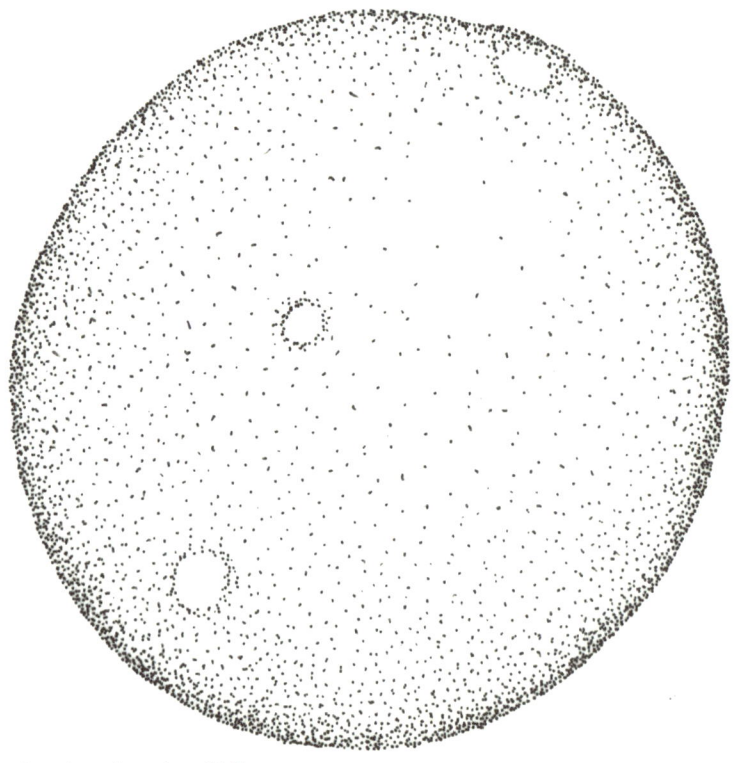

Carpinus betulus 48.0μm
Betulaceae

Another of the birch family, hornbeam is monoecious, having male and female catkins on the same tree.
The pollen grain has four pores around the equator.

Red Horse Chestnut

Large tree, looks very similar to the familiar horse chestnut, but is a hybrid. It is characterised by large, showy red flowers, to attract insects for pollination. The pollen grains are deep red. Note the dots (whatever they are) distributed across the furrows where the pores occur. Polar pollen grain view at the top, equatorial view below right.

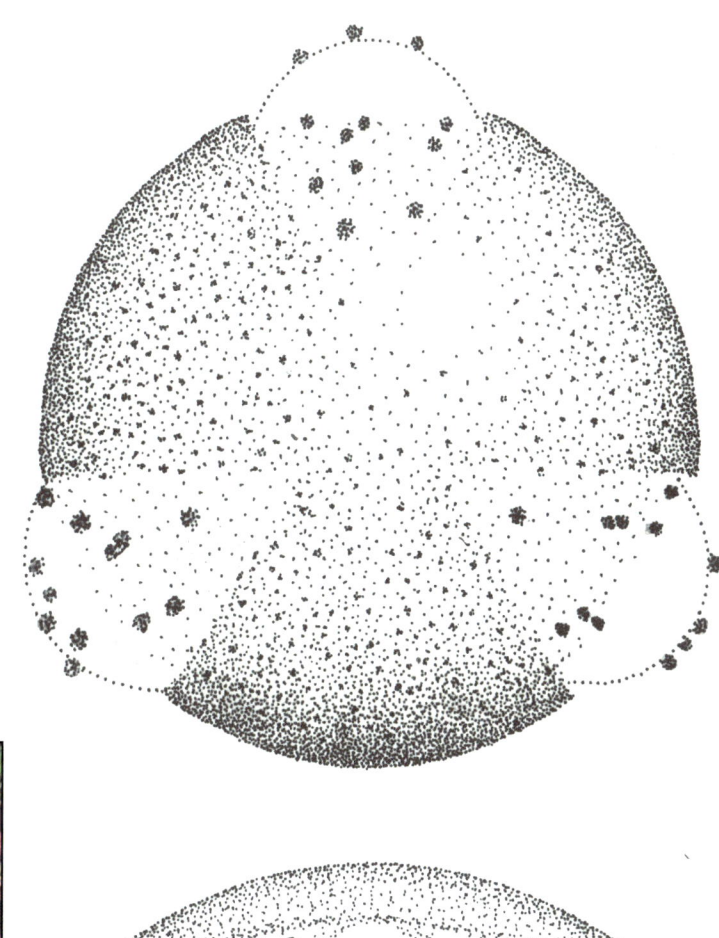

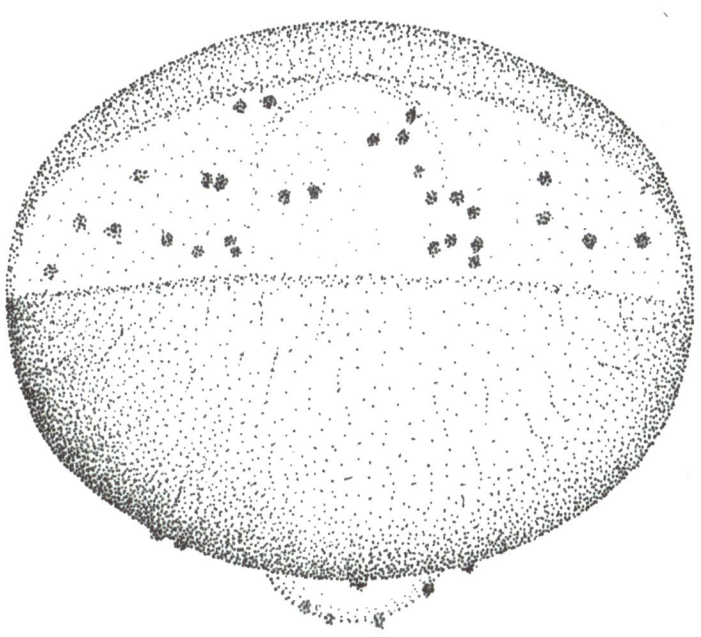

Horse Chestnut (red) Aesculus x carnea 32.5µm
Sapindaceae

Horseshoe vetch

Hippocrepis comosa 17.8µm
Fabaceae

Low growing plant with yellow flowers, each in the form characteristic of the pea family. The flowers are held in an attractive arrangement branching outwards in a circle from the stem.

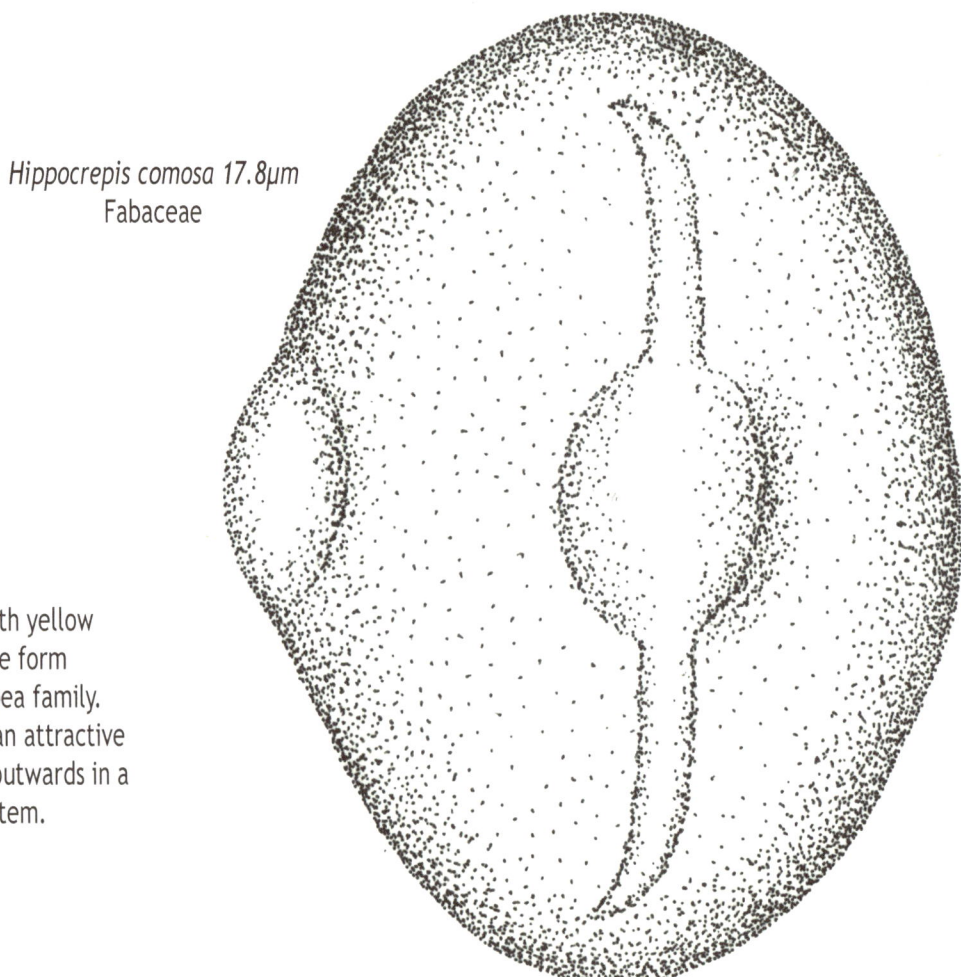

Climbing hydrangea

Hydrangea anomala subsp. *petiolaris* 12.6μm
Hydrangeaceae

Climbing hydrangea with curious flat flower heads with great numbers of small, off-white fertile flowers at the centre and smaller numbers of larger sterile flowers around the edges. Polar view of the pollen grain showing the three furrows almost meeting at the pole.

Ice plant

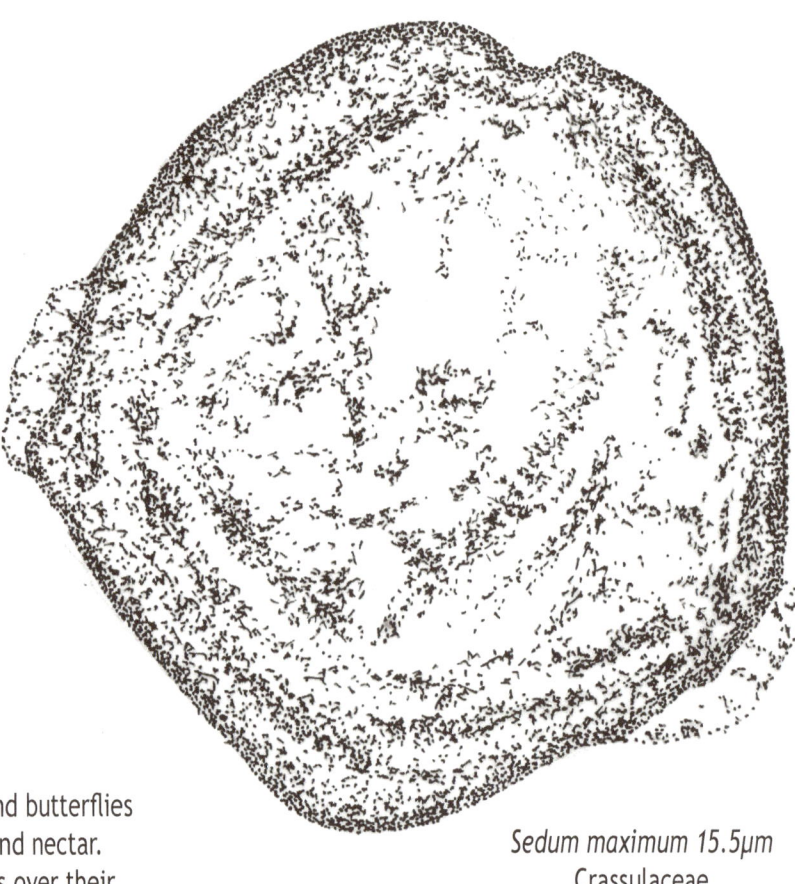

A late flowering plant attractive to bees and butterflies giving a good supply of autumn pollen and nectar. Sedum pollens have a latticework of lines over their surface, and quite prominent pores, even when dry.

Sedum maximum 15.5µm
Crassulaceae

Ground ivy

Photos © Sally Dunn

Glechoma hederacea 56.0μm
Lamiaceae

Polar and equatorial views showing the six furrows.

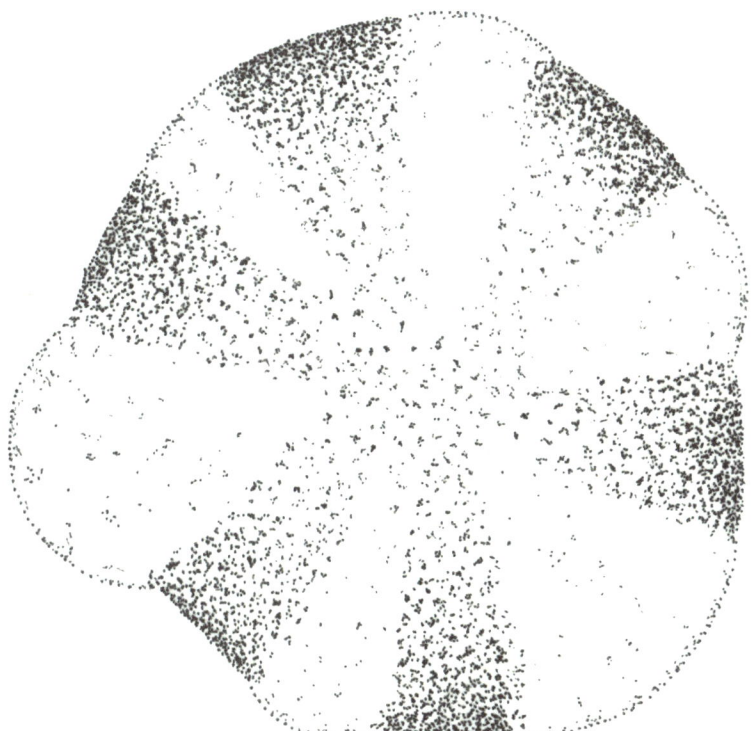

Ivy

The last major source in the year for both pollen and nectar, though the high glucose to fructose ratio in ivy nectar can cause honey to crystallise in the comb, rendering the stores inaccessible to honey bees during winter.

The illustration, immediately below, shows the wide latticework over the surface and three furrows which widen or close depending on humidity to reveal or protect the pores at their equators.

Photo © Sally Dunn

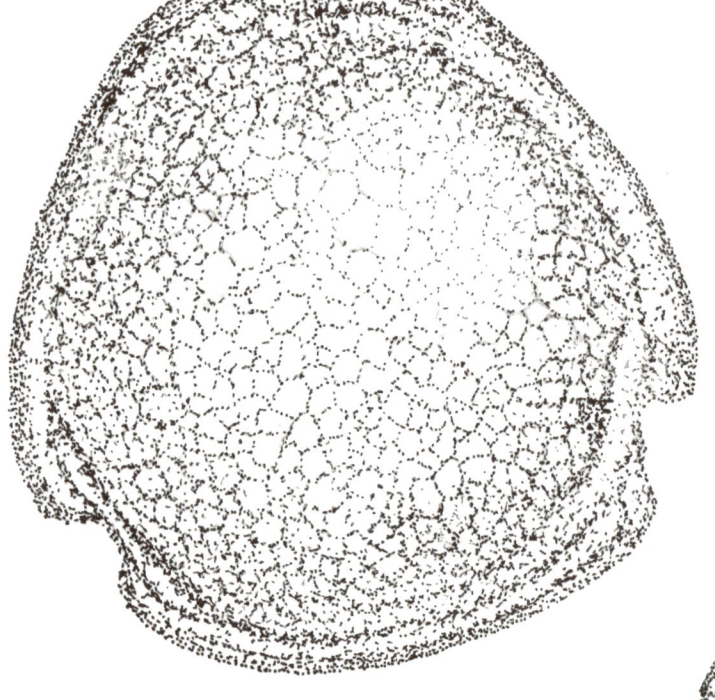

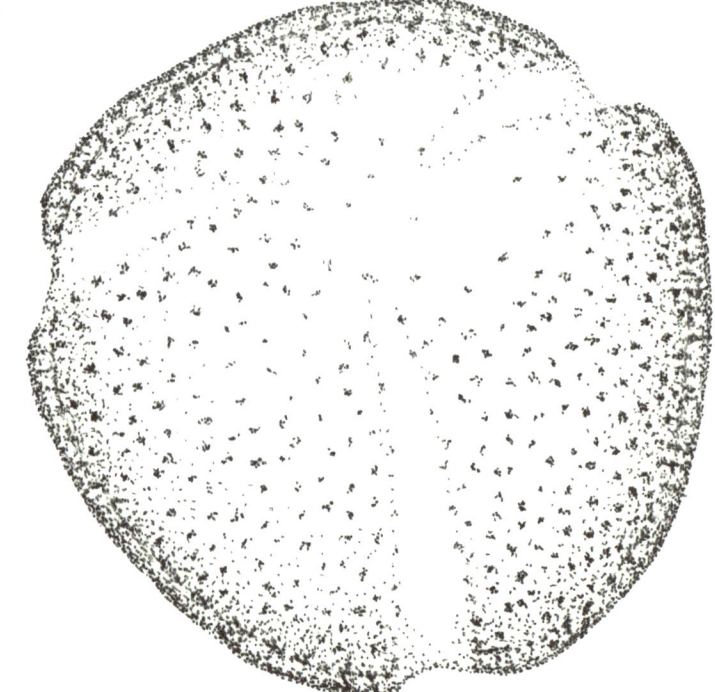

Hedera helix cross section above, 26.2µm polar view right 26.1µm
Araliaceae

Japanese quince

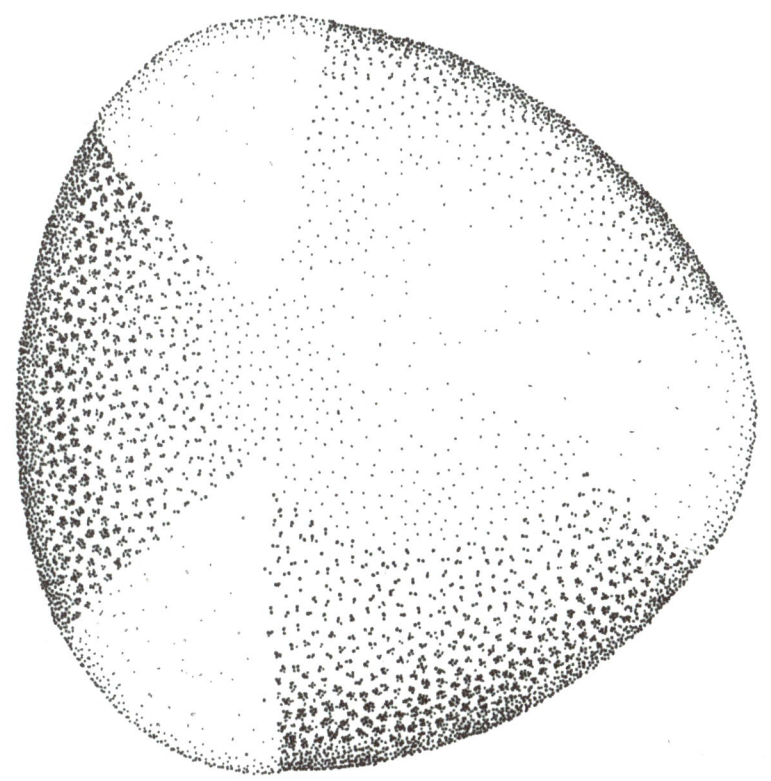

Chaenomeles speciosa 40.0μm
Rosaceae

An early flowering shrub related to the quince (*Cydonia oblonga*) grown ornamentally for its flowers. The pollen grain has a common formation of three furrows.

Laburnum

Laburnum anagyroides 25.0µm
Fabaceae

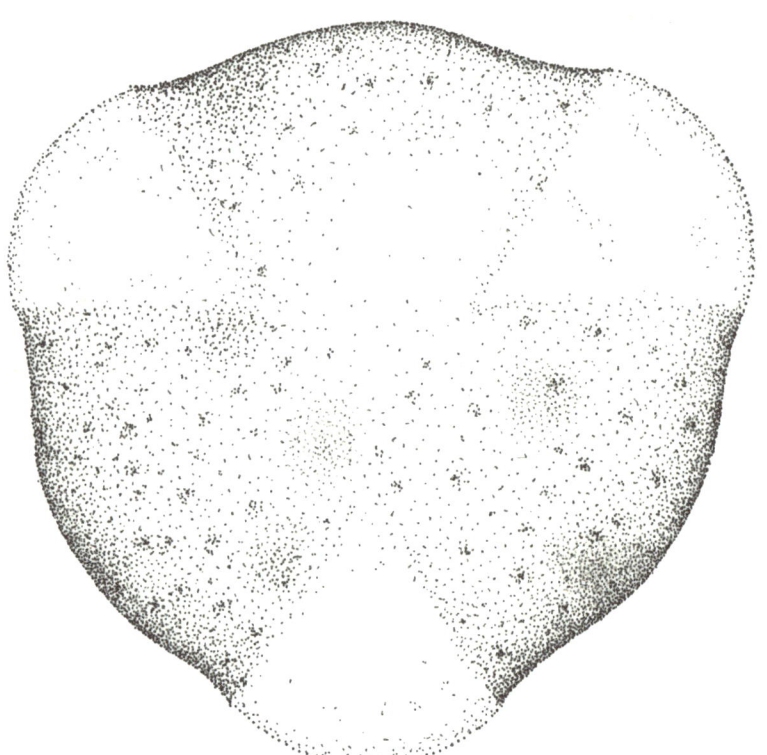

A medium sized tree popular for the stunning display of yellow racemes in spring. It also attracts bees for pollination.
The pollen grain has a close patterned surface and three wide furrows.

Lady's bedstraw

A herbaceous perennial wild flower named after one of its uses. The pollen has six furrows.

Galium verum 20.0µm
Rubiaceae

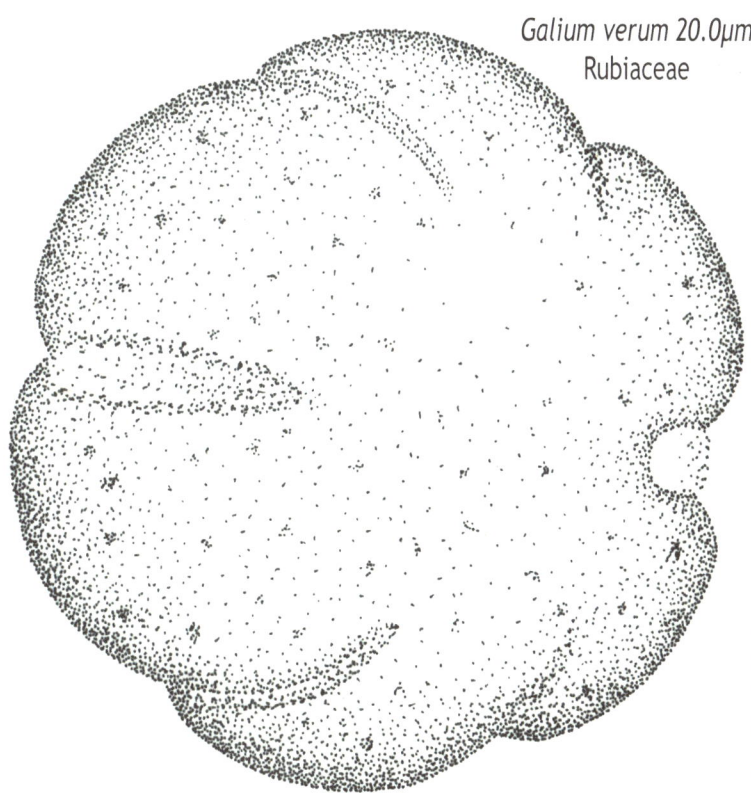

Cherry Laurel

Photos © Sally Dunn

An evergreen shrub widely used for hedging producing long, upright, scented racemes of hermaphrodite flowers in spring, followed by black 'cherries' in autumn. It is pollinated by insects, its younger leaves having extrafloral nectaries on their undersides attracting honeybees.

Prunus laurocerasus 37.5μm
Rosaceae

Lavender

Alive with bees when in flower, pollen from the tiny, tightly closed flowers is not the easiest to extract for microscopy.
The pollen grain is among a small number having six furrows. Its exine is peppered with holes.

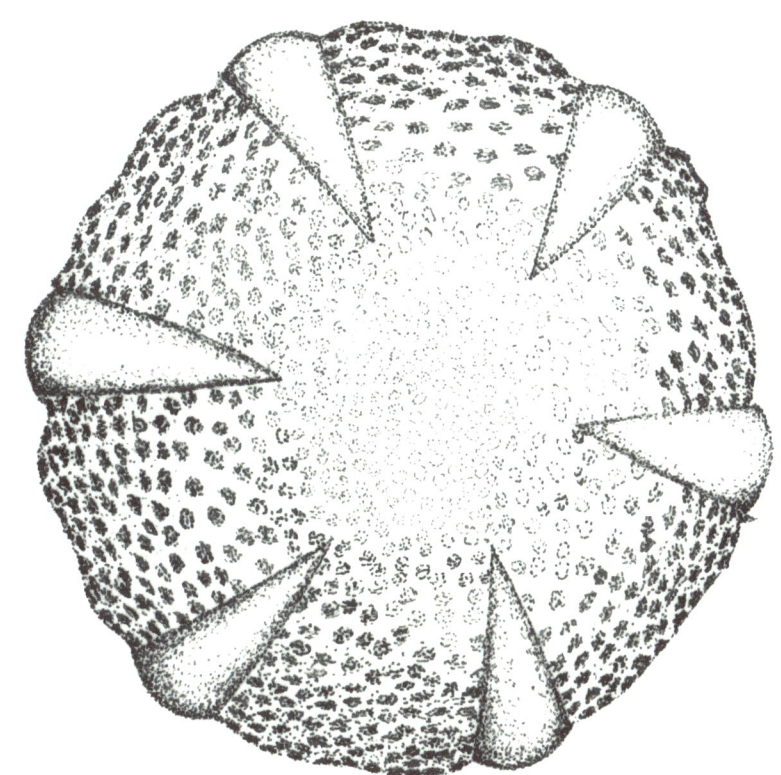

Lavandula angustifolia 37.8µm
Lamiaceae

Leptospermum myrtifolium

Leptospermum myrtifolium 17.5µm
Myrtaceae

The leptospermum family, native to Australia and New Zealand, are frost hardy to a degree in the UK. Shrubby drought resistant evergreen bush with silvery leaves.
The pollen grain is small and roughly triangular in shape.

Leptospermum scoparium (Manuka)

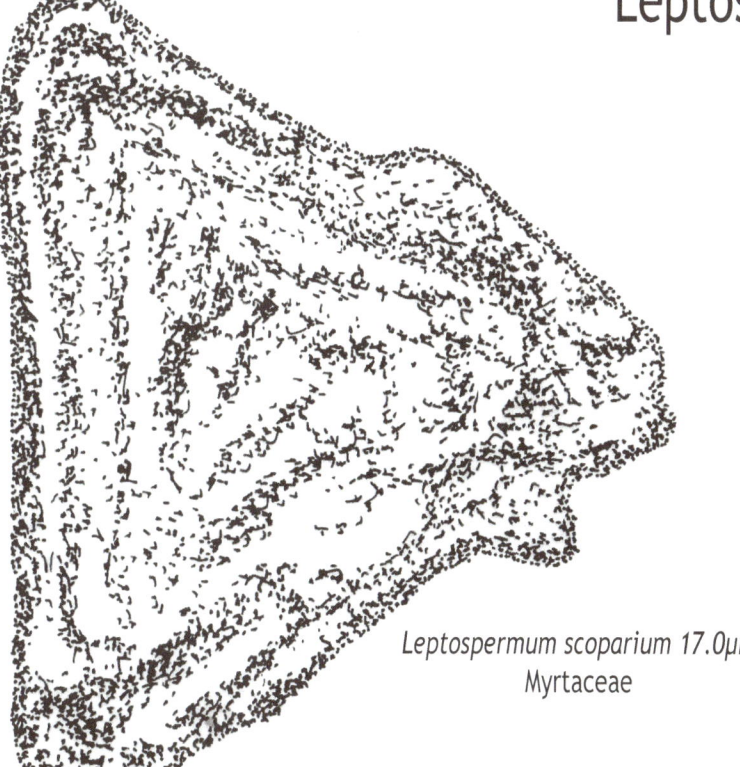

Leptospermum scoparium 17.0μm
Myrtaceae

The leptospermum family prefer the acid soil and milder climate in parts the south west of the UK where they can grow to 12 or 15 feet high. There are cultivars with various attractively coloured flowers from white to deep pink and red.
Nectar and hence honey from *Leptospermum scoparium* fetches a premium as it prized for reputed medicinal qualities.
The pollen grain is triangular with unequal sides.

Photo © Stephen Rhenius

Lilac

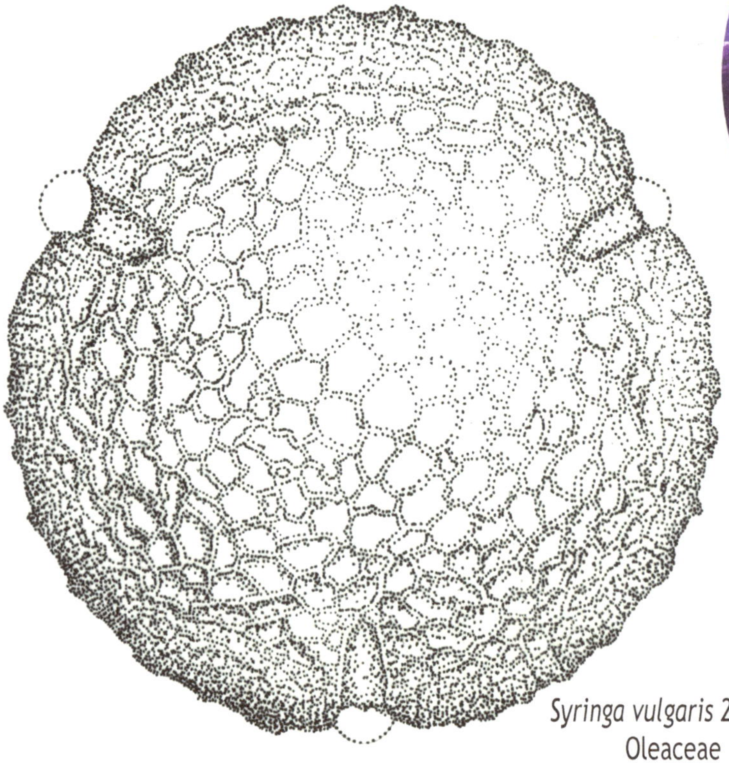

Syringa vulgaris 26.2μm
Oleaceae

The strongly scented flowers produce both pollen and nectar. Note the apertures in the net like pattern on the exine.

Photo © Sally Dunn

Lime

Large trees, often lining avenues and sweeping country driveways. Flowering in June - July, the hermaphrodite flowers are pollinated by insects. Bees foraging on lime produce a golden, aromatic honey. The pollen grain shape is oblate so tends to lie flat with the view under the microscope usually a polar view.

Tilia sp. 39.9µm
Malvaceae

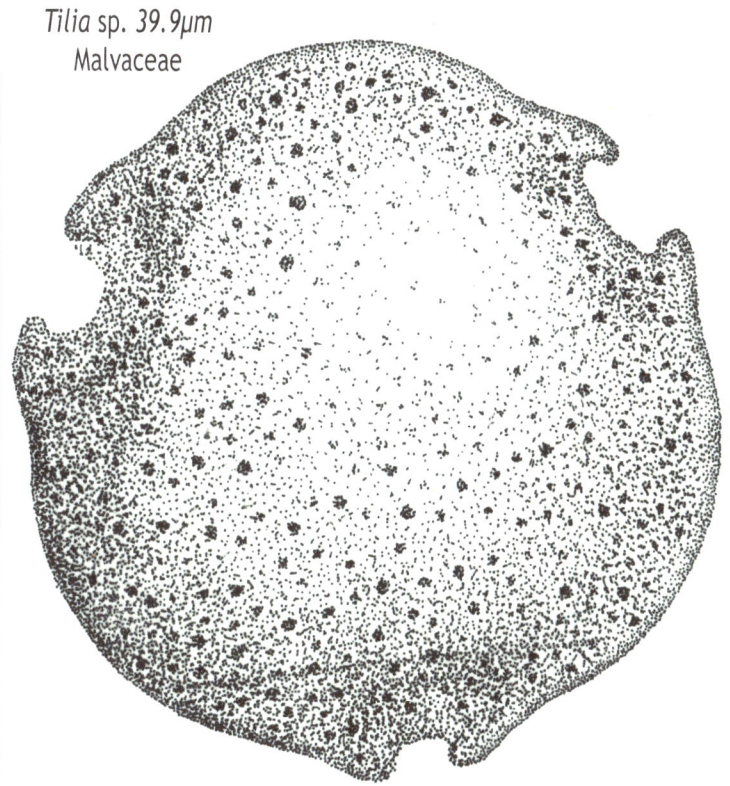

Limnanthes douglasii

The poached egg plant, also known as meadowfoam, attracts hoverflies as well as bees.
The pollen grain, shown here in equatorial view, has one furrow which completely circles the grain.

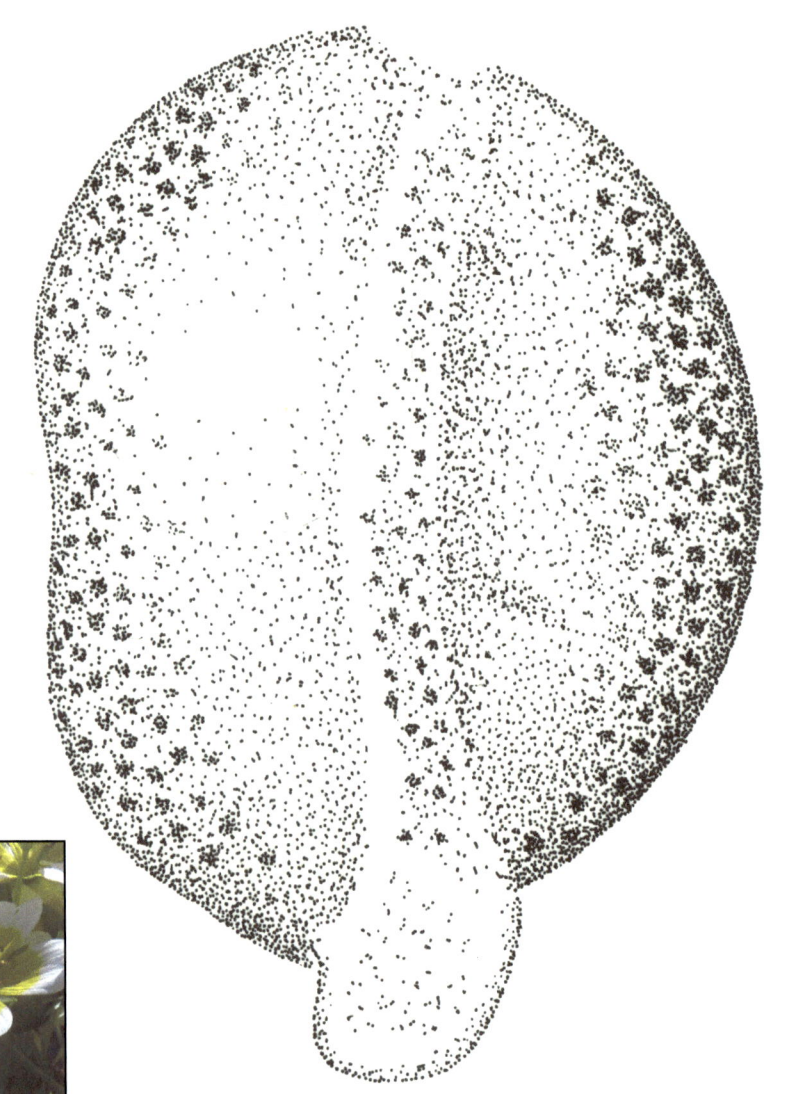

Limnanthes douglasii 25.0µm
Limnanthaceae

Lobelia

Much loved annual summer bedding plant for flower beds, containers and baskets. The delicate flowers are bent right over when visited by heavy bumble bees.
The pollen has three furrows which almost meet at the poles.

Lobelia erinus 23.9µm
Campanulaceae

London Pride

The sexine (sculptured exine) pattern of many of the Saxifragaceae has been studied[9], of which the morphological types and subtypes are generally found to support existing classifications. This drawing shows the striated surface pattern.

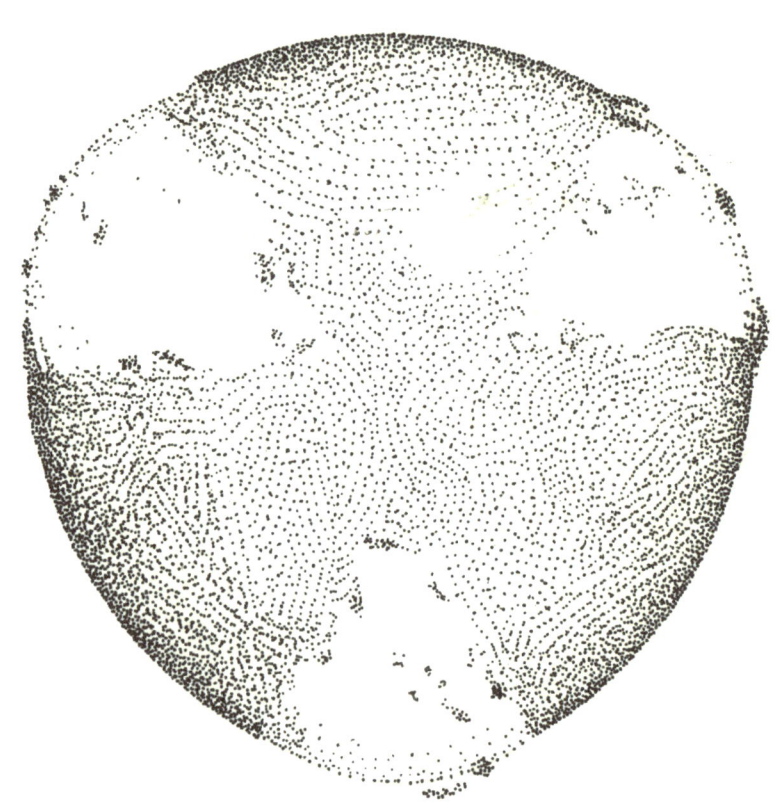

Saxifraga × urbium 37.0µm
Saxifragaceae

Loosestrife (purple)

Tall herbaceous perennial, usually found near water, with square, reddish stems and lanceolate leaves.
The pollen grain notably a striated surface pattern and six furrows. The furrows alternate in proximity to each pole.

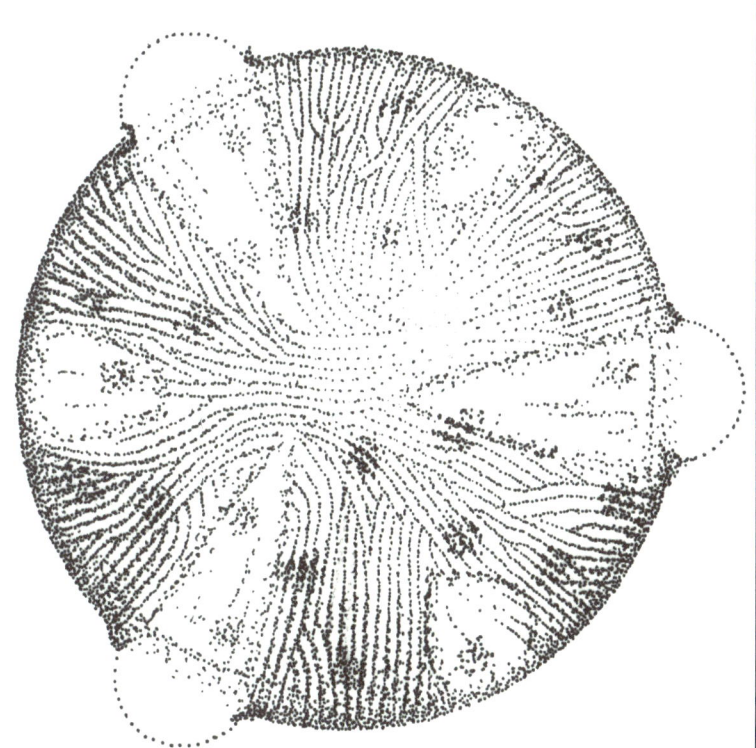

Lythrum salicaria 43.0µm
Lythraceae

Lords-and-Ladies

A woodland plant with distinctive flame shaped flower structure in spring. The flowers are at the base of the spadix with a ring of female flowers at the bottom and a ring of male flowers above them. The pollen is disappointingly not at all distinctive. They are insect pollinated, producing a scent attractive to small insects. Their fruit, in autumn. consists of sprigs of conspicuous, bright red, poisonous berries.

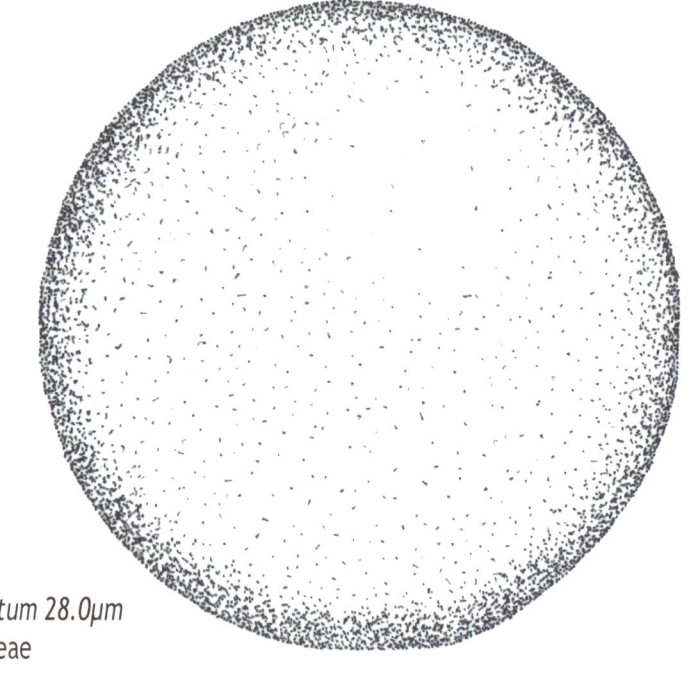

Arum maculatum 28.0µm
Araceae

Inside, at the base of the structure, the spherical female globes - which later develop into the red berries - are pollinated by small flies which collect pollen as they brush across the reddish brown male flowers. The insects are then trapped inside the strucure by the ring of hairs above and enclosed by the bract.

Lucerne

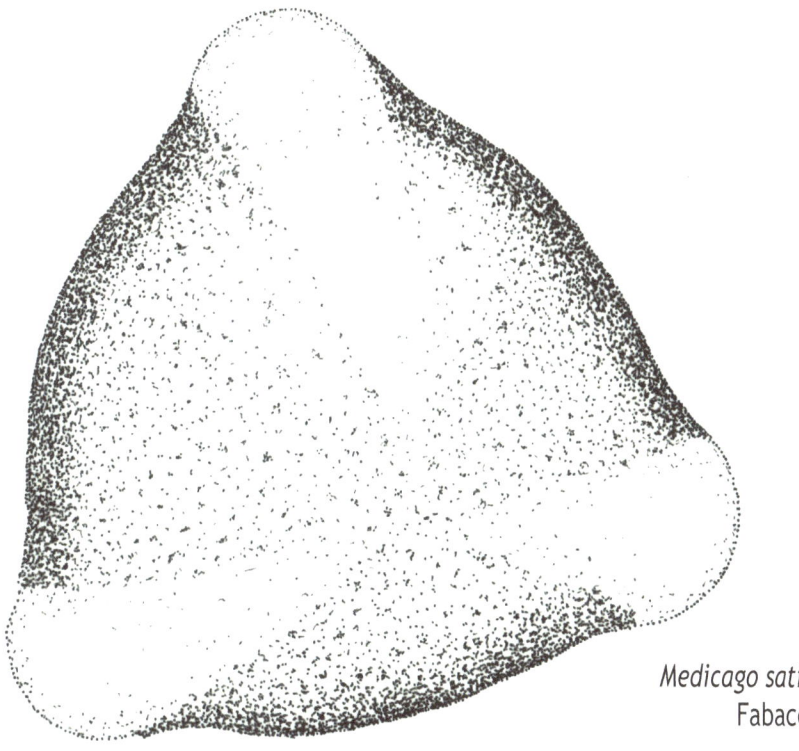

Useful forage crop both for insects and animals.
Both views shown here are polar with different shading techniques demonstrating shape of the grain and (below) the reticulated surface of the pollen grain.

Medicago sativa 32.0µm
Fabaceae

Photo © Janet Morris

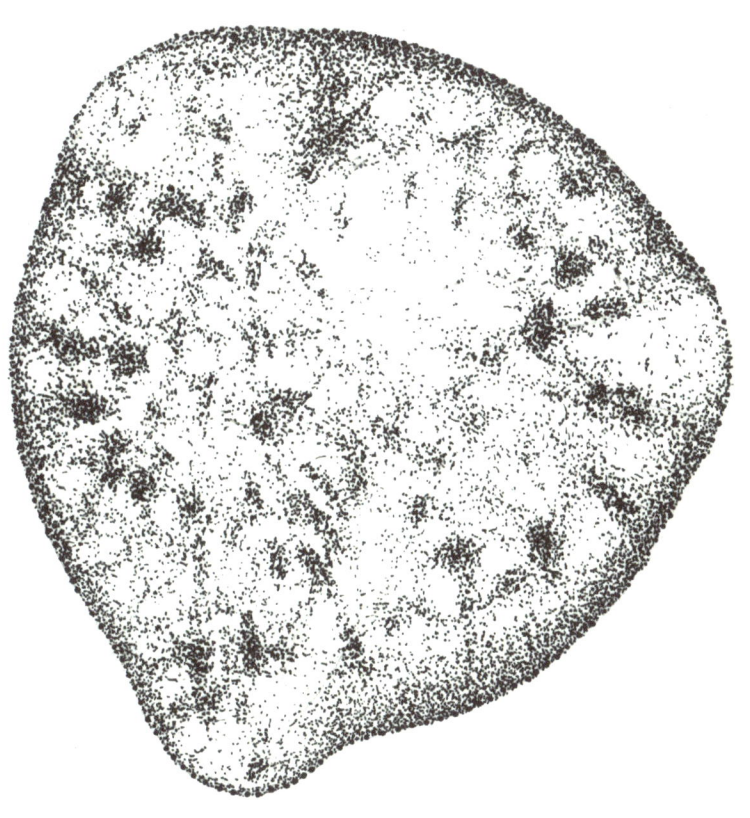

Mahonia

Photo © Janet Morris

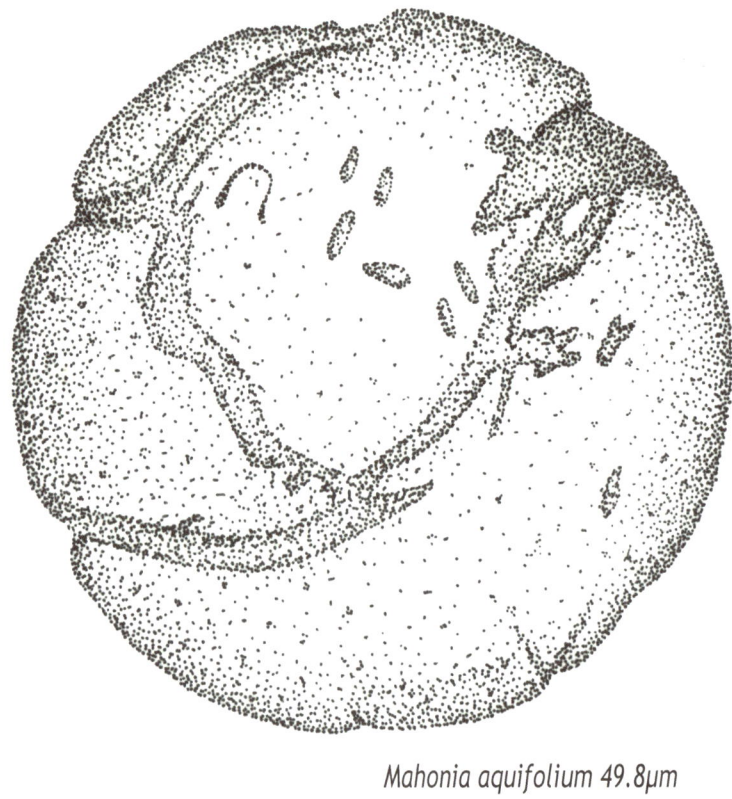
Mahonia aquifolium 49.8µm
Berberidaceae

Flowers are borne on spreading racemes. There are many varieties flowering from mid winter onwards, a magnet for winter pollinators. Mahonia stamens are sensitive. When touched by a bee looking for nectar they spring inwards and deposit pollen on the bee's head. The exine on the pollen grain seems fragile, as though breaking apart. It has a lattice work of furrows dividing the surface into pentagonal shapes.

Maize

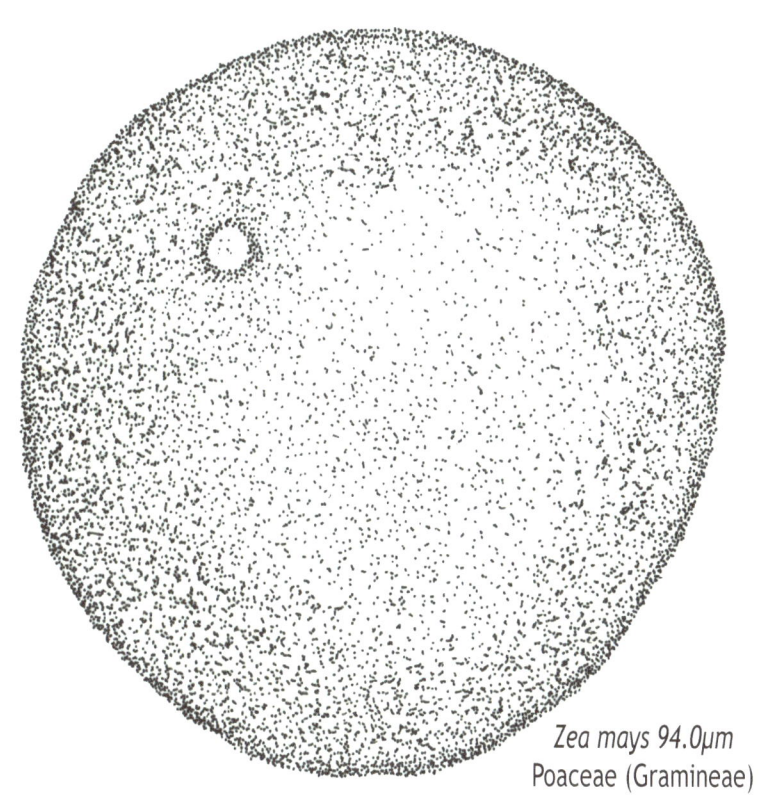

Zea mays 94.0μm
Poaceae (Gramineae)

In common with the grass family, maize pollen, though quite large, has just one pore. The plant is tall with the pollen shedding male flowers at the top, and the female flower just visible two thirds of the way down the right hand side of the plant. The female flower will develop into the corn cob.

Common Mallow

One of the larger pollen grains, spherical, with a spiny surface and tiny holes, rather than obvious pores.

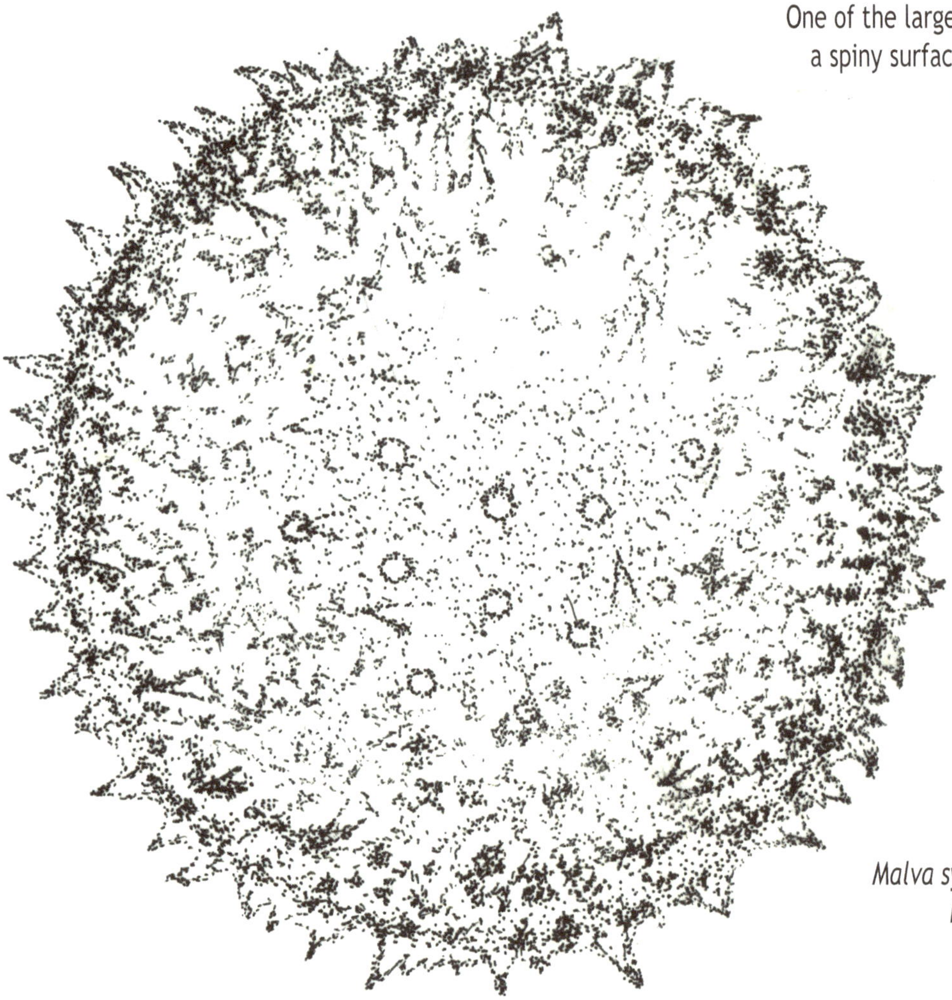

Malva sylvestris 109.0μm
Malvaceae

Marjoram

A culinary herb attractive to insects.
The pollen grain has six furrows.

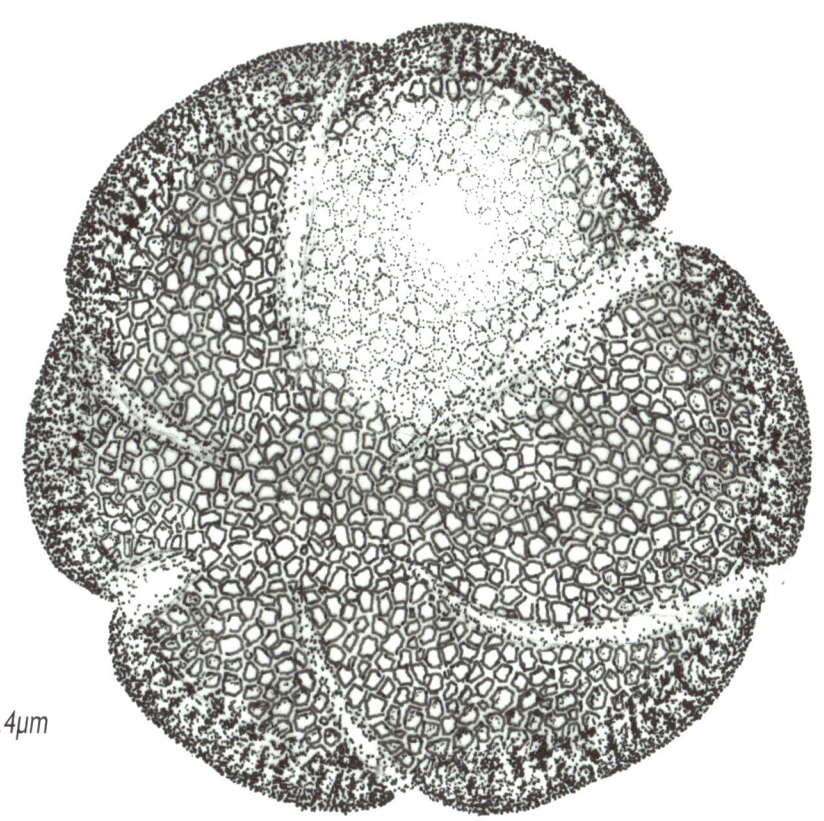

Origanum vulgare 35.4μm
Lamiaceae

Marrow

Large flowers with easy access for insects, and huge pollen grains with visible pores.

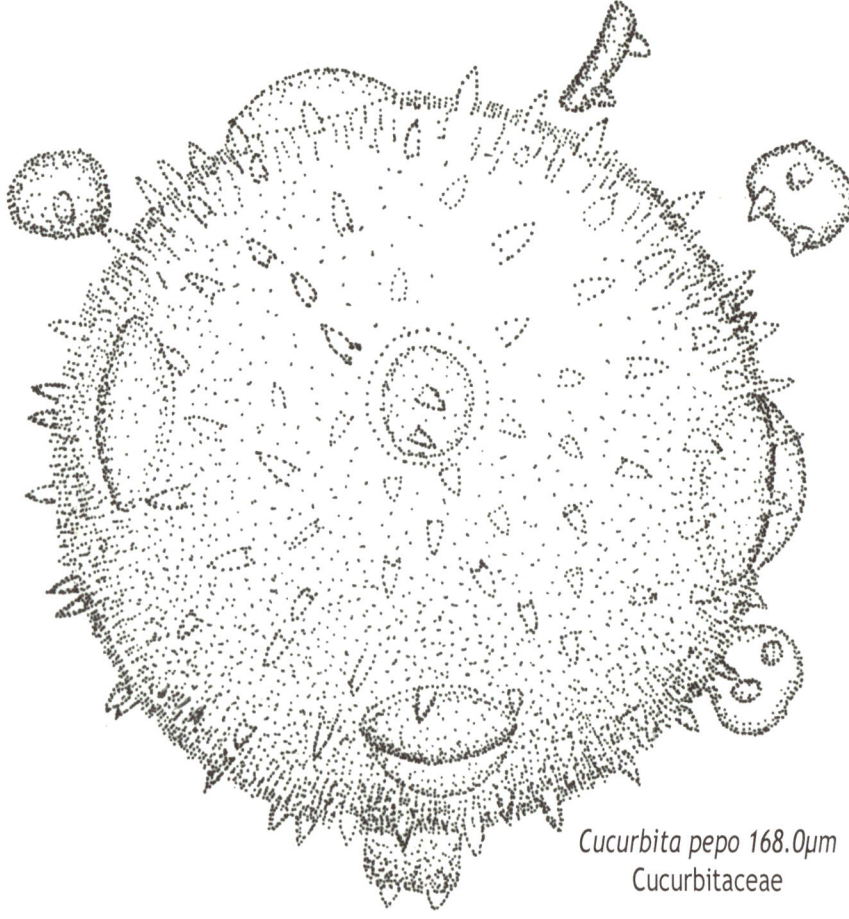

Cucurbita pepo 168.0µm
Cucurbitaceae

Meadowsweet

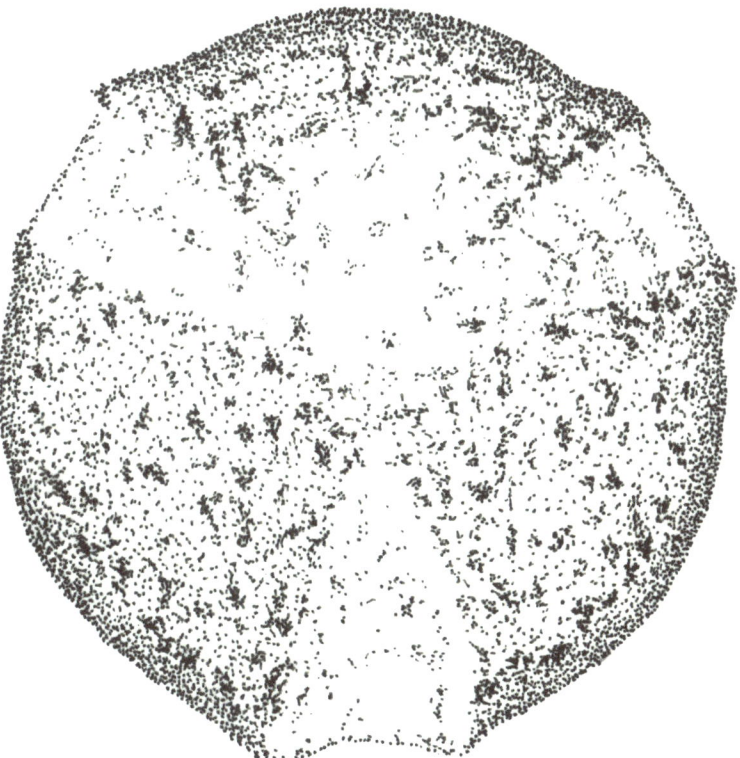

Filipendula ulmaria 17.3µm
Rosaceae

Perennial herbaceous plant with
strongly scented frothy flowers visited
by various insects,
particularly flies.
Small pollen grain with three furrows.

Ribbed melilot

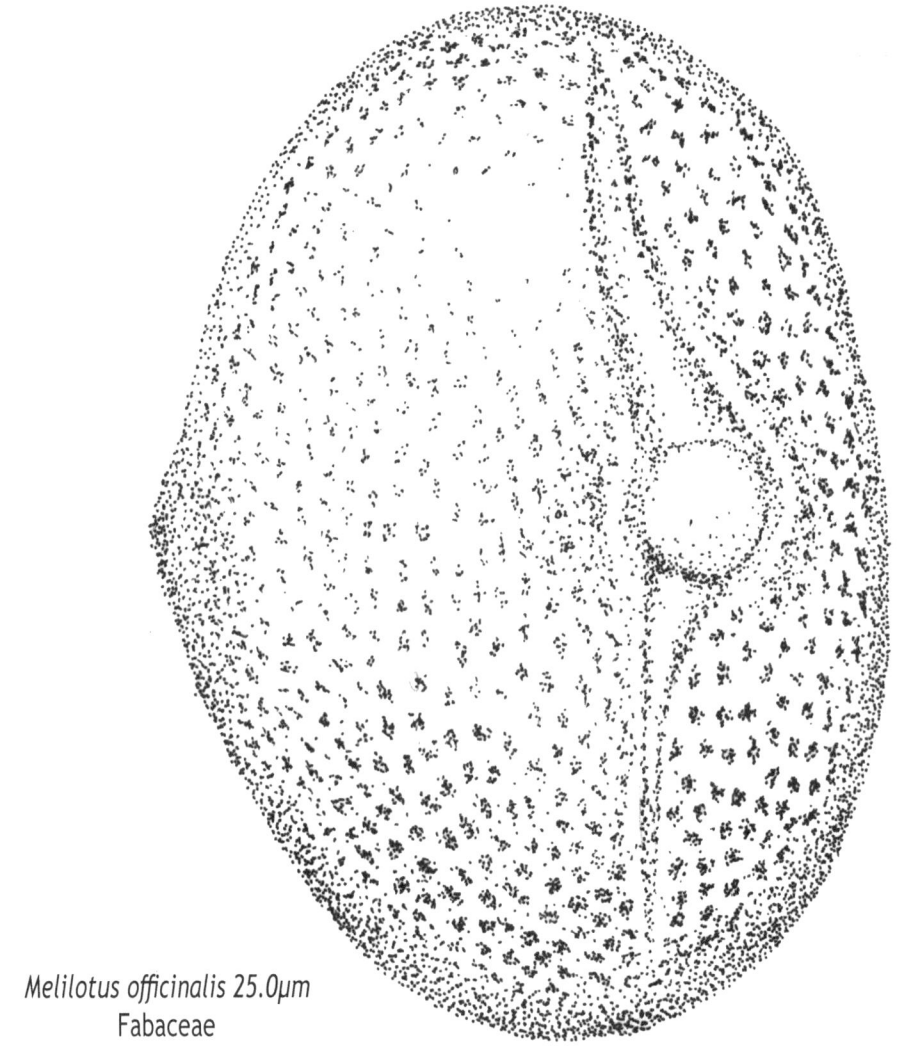

Melilotus officinalis 25.0μm
Fabaceae

Also known as sweet clover, ribbed melilot has been used to improve nitrogen content in poor soils. It is a major source of nectar for honey bees resulting in high honey yields.
The pollen grain shape is prolate or elongate so lies on its side with the view under the microscope always an equatorial view.

Annual Mercury

Male fllowers are borne on spikes, female flowers in small clusters.
Small pollen grain with three decorated furrows.

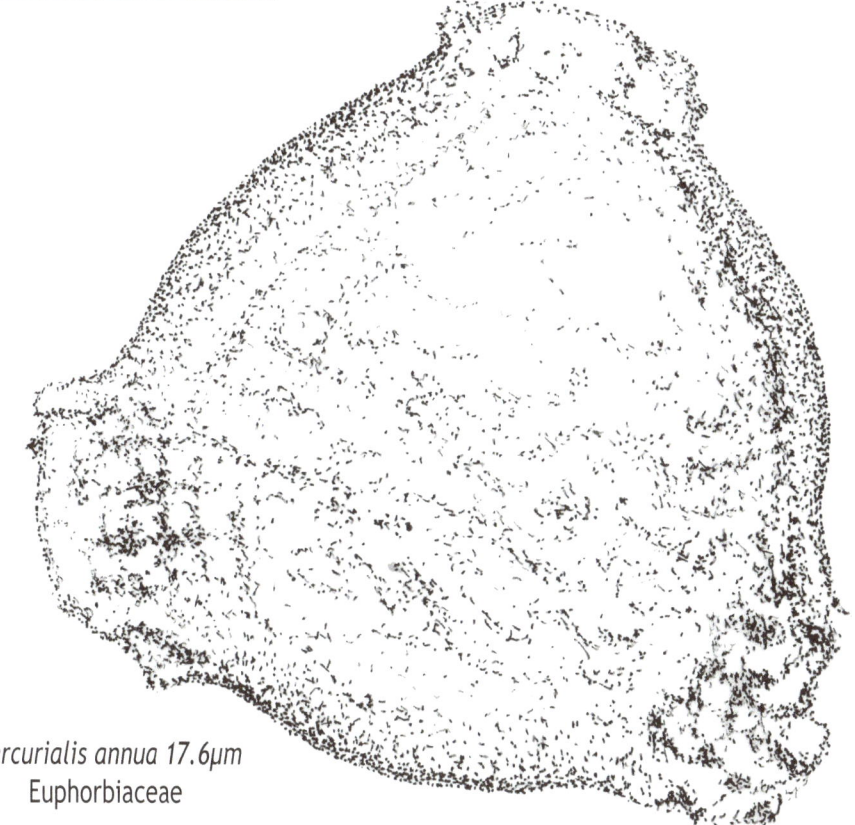

Mercurialis annua 17.6μm
Euphorbiaceae

Michelmas daisy

Symphyotrichum novi-belgii (Aster novi-belgii) 30.0μm
Asteraceae (Compositae)

Good source of autumn pollen, the grains having the usual spikes, arranged in neat whorls, and three prominent pores.

Mimosa

A large tree with curious yellow pom pom flower heads. Mimosa pollen has an unusual tetrahedron formation. The four central grains are two deep, totalling sixteen clumped together like a small cushion.
The grains have no apertures.

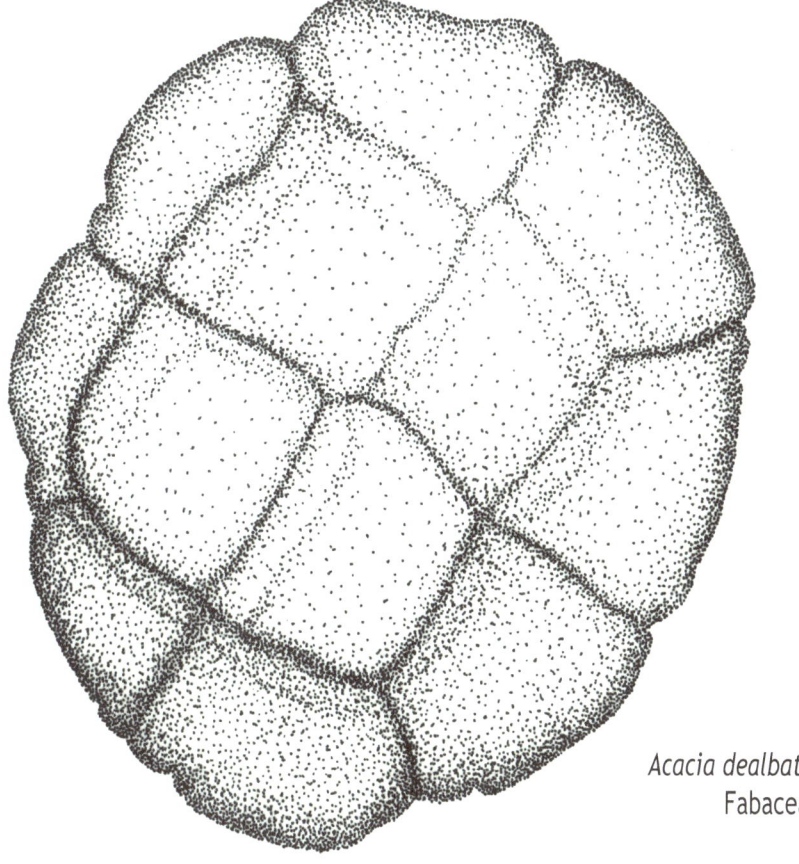

Acacia dealbata 44.4μm
Fabaceae

Mint (water)

Herbacious perennial,
the pollen grain having six furrows.

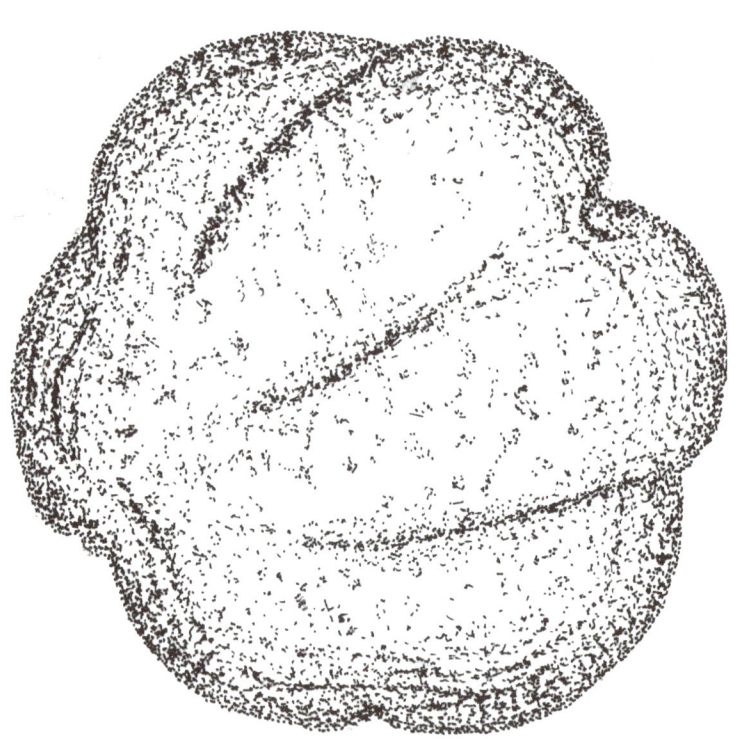

Mentha aquatica 34.2µm
Lamiaceae

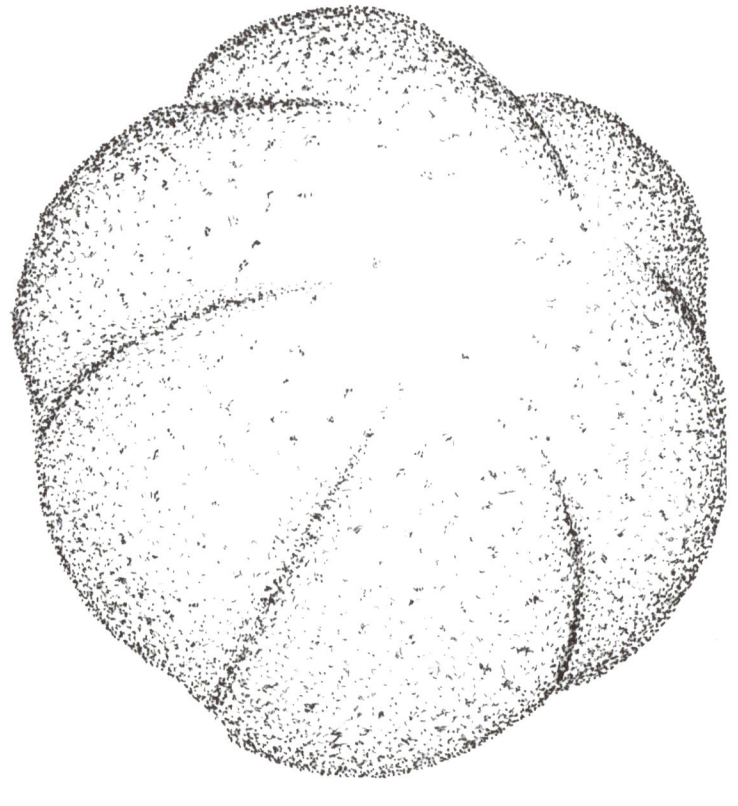

Mistletoe

Photo © Sally Dunn

A parasitic plant which grows on the branches of trees and has a spherical outline, unlike untidy birds' nests. The male plants produce pollen, and female plants produce the berries for which mistletoe is famous.

The pollen grain's three furrows are flush with the surface of the rest of the grain, with a finer, denser pattern of lower spikes covering them.

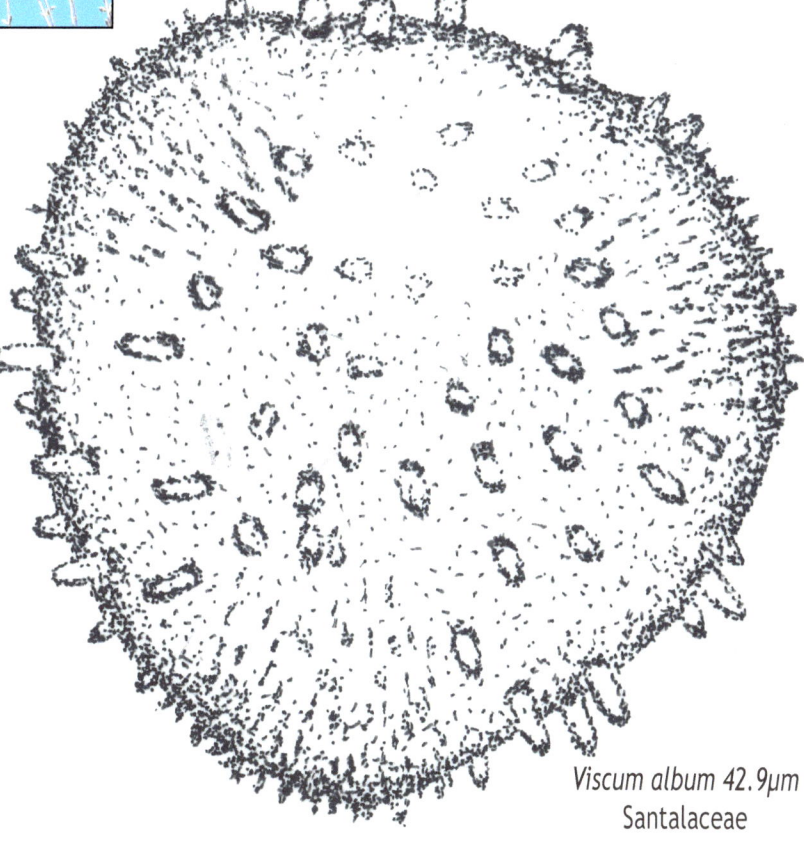

Viscum album 42.9µm
Santalaceae

Morning Glory

Ipomoea tricolor 147.0µm
Convolvulaceae

An attractive annual garden climber. Like its relative, bindweed, a large pollen grain with spikes and visible pores dotted over the surface.

Mugwort

Hermaphrodite, having both male and female organs and attractive to wildlife. The plant flowers have no petals. This suggests that it is wind pollinated. The drifting pollen can cause late summer hay fever.
The furrows are recessed. This is one of few pollen grains, shown in polar view, among the daisy family that has no broad based spines.

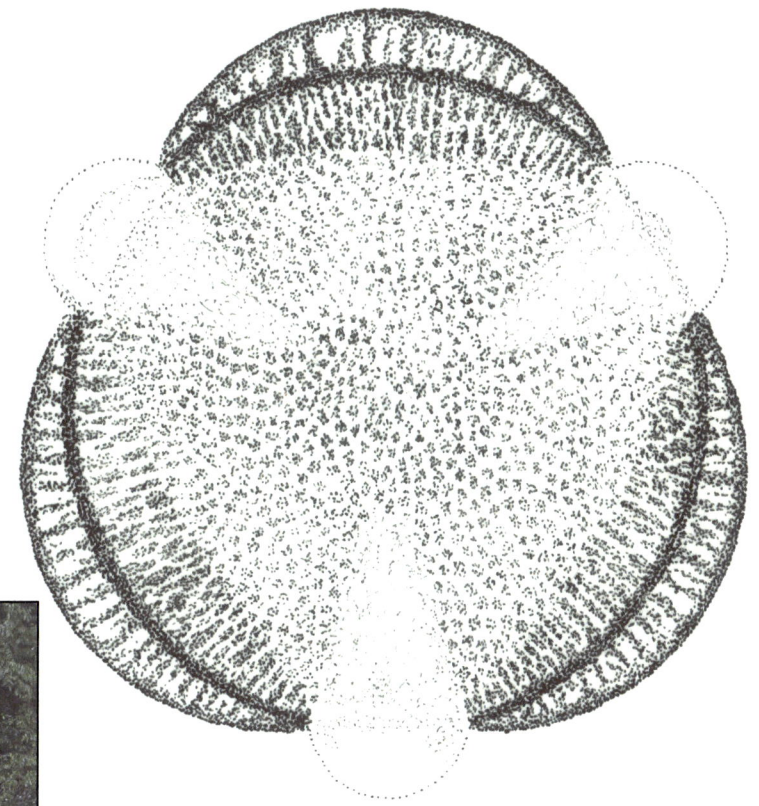

Artemesia vulgaris 26.2µm
Asteraceae (Compositae)

Great mullein

Tall, biennial wild flower with visibly reddish orange anthers. The pollen grain has a surface netting but the three furrows have a smooth surface.

Verbascum thapsus 22.7μm
Scrophulariaceae

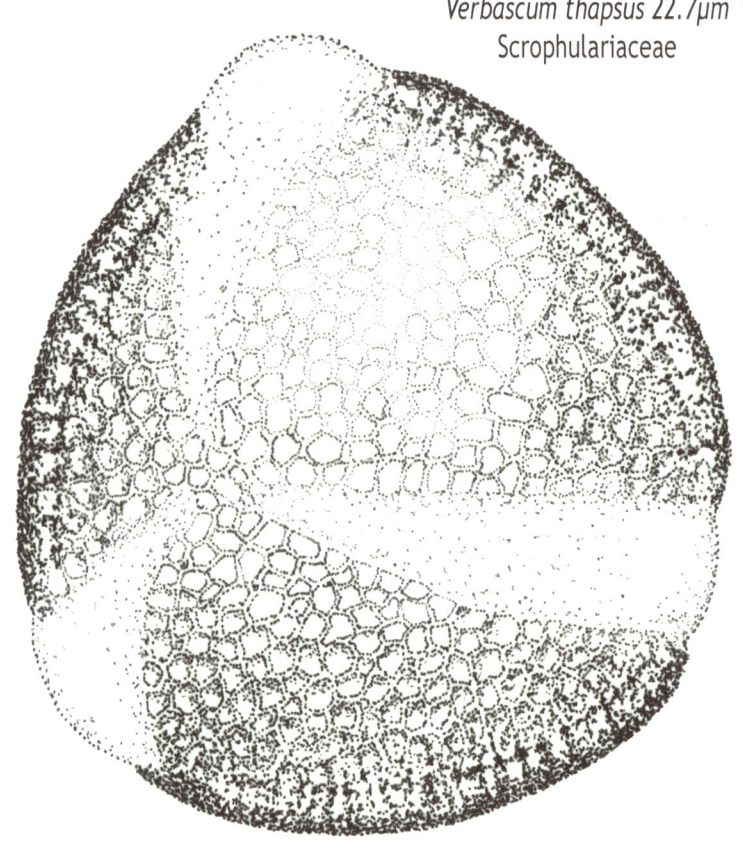

Nasturtium

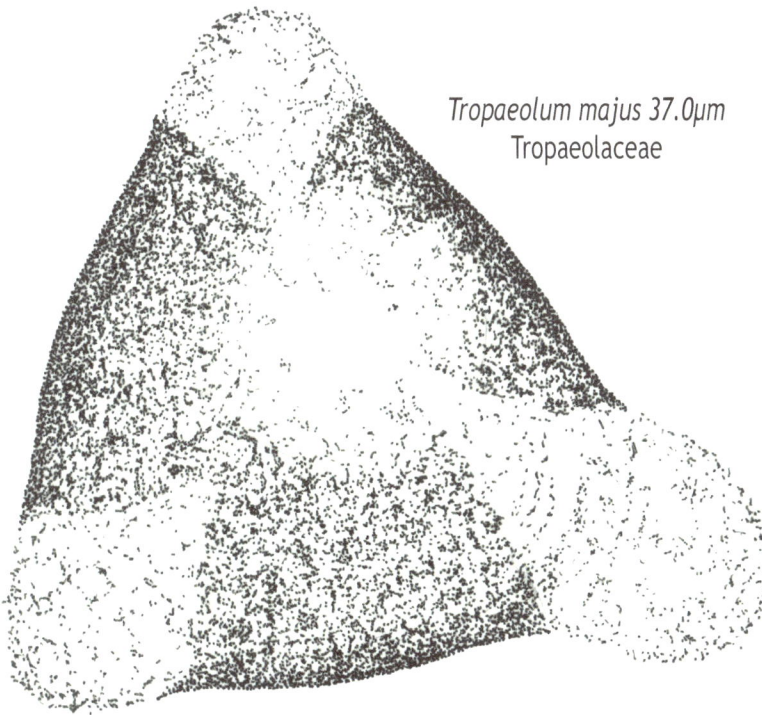

Tropaeolum majus 37.0µm
Tropaeolaceae

Popular, edible, annual with showy flowers, thriving in poor soil. The sucrose rich nectar, deep in the spur, can only be reached by a long tongued insect.
The surface of the grain is latticed.

Red dead-nettle

Another favourite for visiting bees, its pollen is red coloured. The pollen grain has a very small polar field.

Photo © Sally Dunn

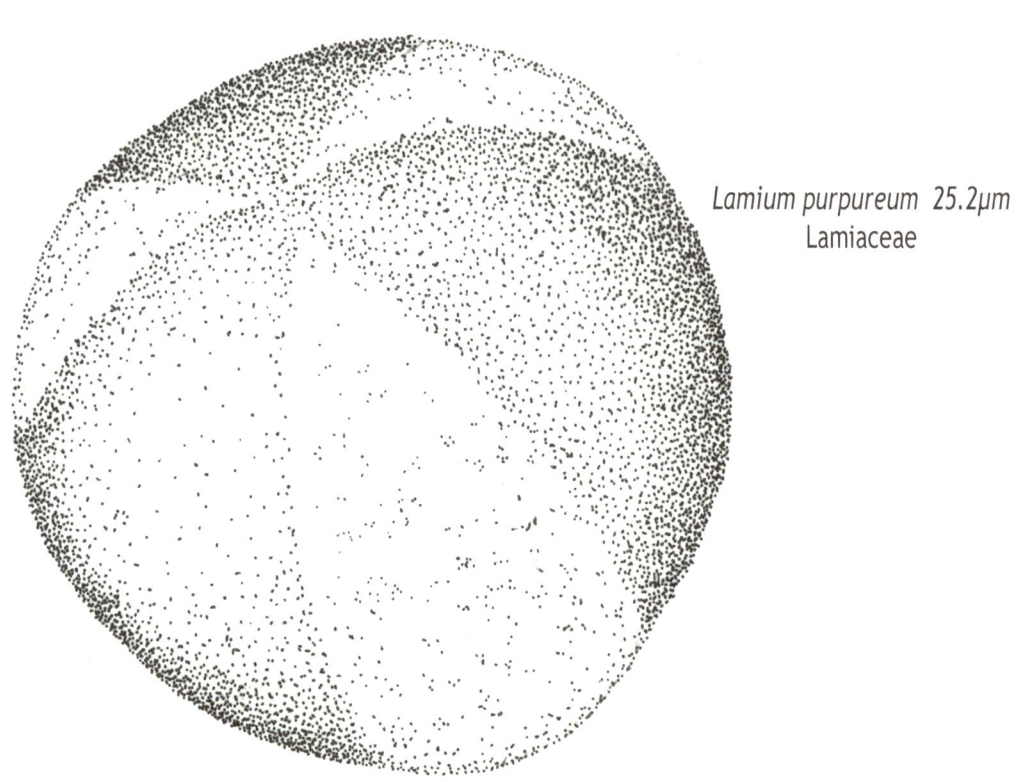

Lamium purpureum 25.2µm
Lamiaceae

Stinging nettle

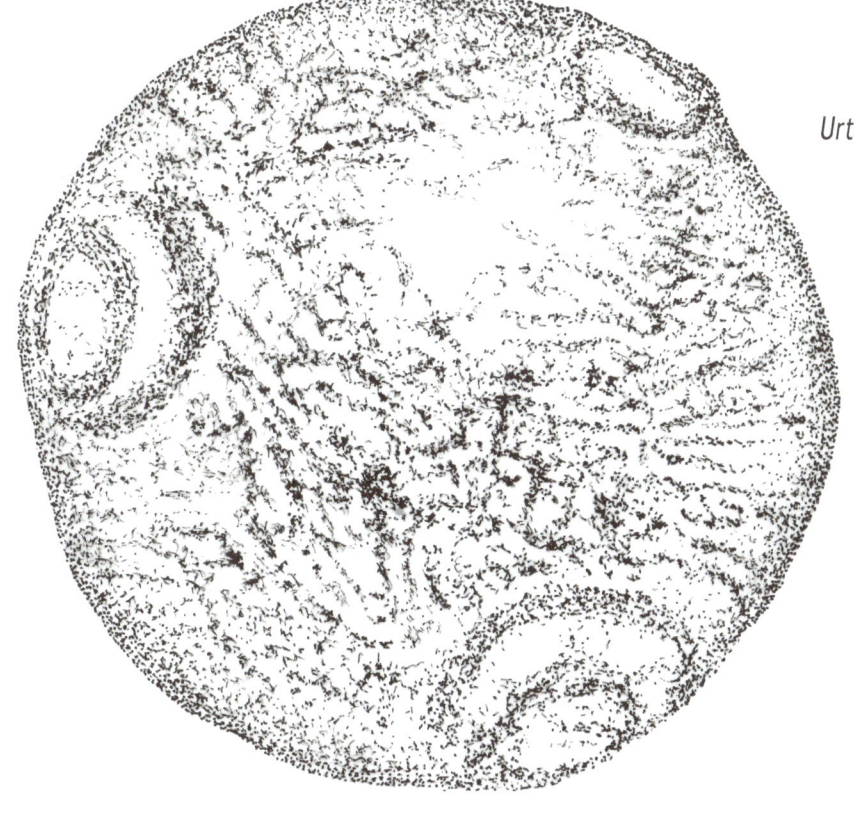

Urtica dioica 20.3µm
Urticaceae

The stinging nettle grows best in rich soil and is itself a good garden fertiliser. It is a host to a huge number of insect species being the larval food plant for, among others, the peacock, comma and red admiral butterflies. These in turn attract other wildlife.
The stinging nettle has a spherical pollen grain with a textured surface and three prominent pores.

White dead-nettle

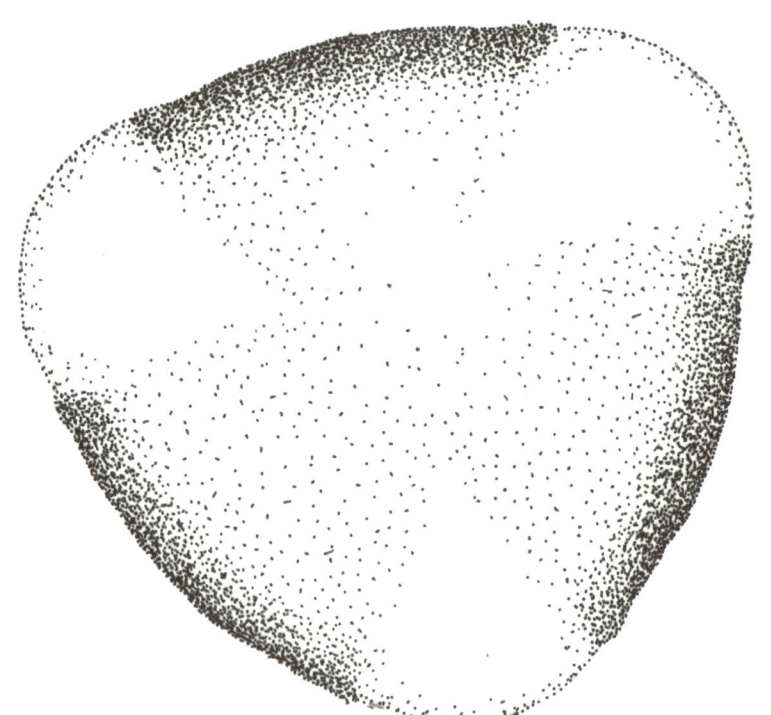

Lamium album 22.6μm
Lamiaceae

Common wild flowering plant, a great
favourite with bumble bees.
The pollen grain has a densely woven
pattern over most of its surface.

Oak

Quercus robur 25.0µm
Fagaceae

Unobtrusive catkins and fairly small, smooth surfaced pollen grain, characteristic of wind pollinated trees.

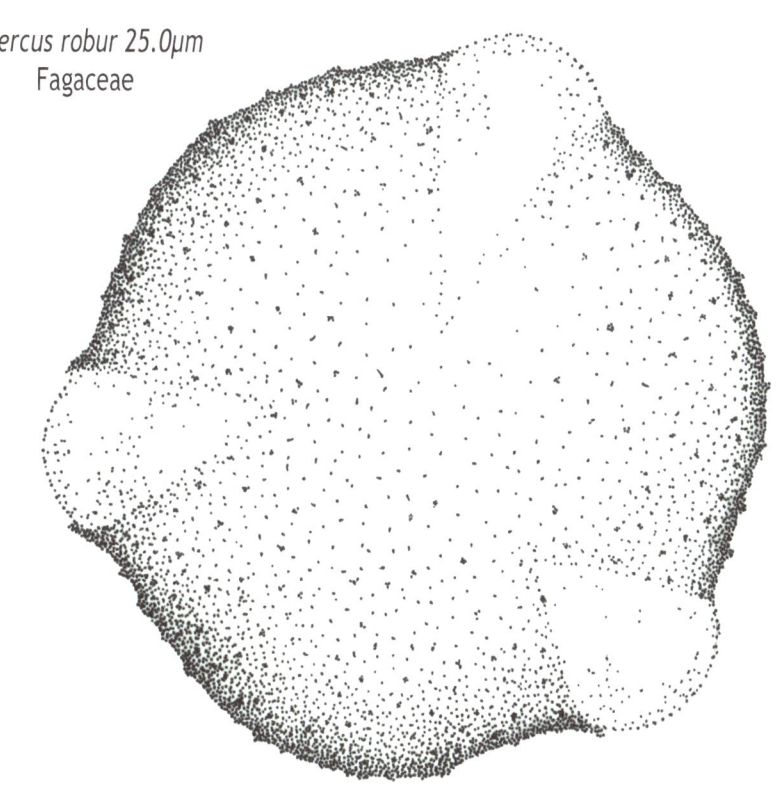

Oil seed rape

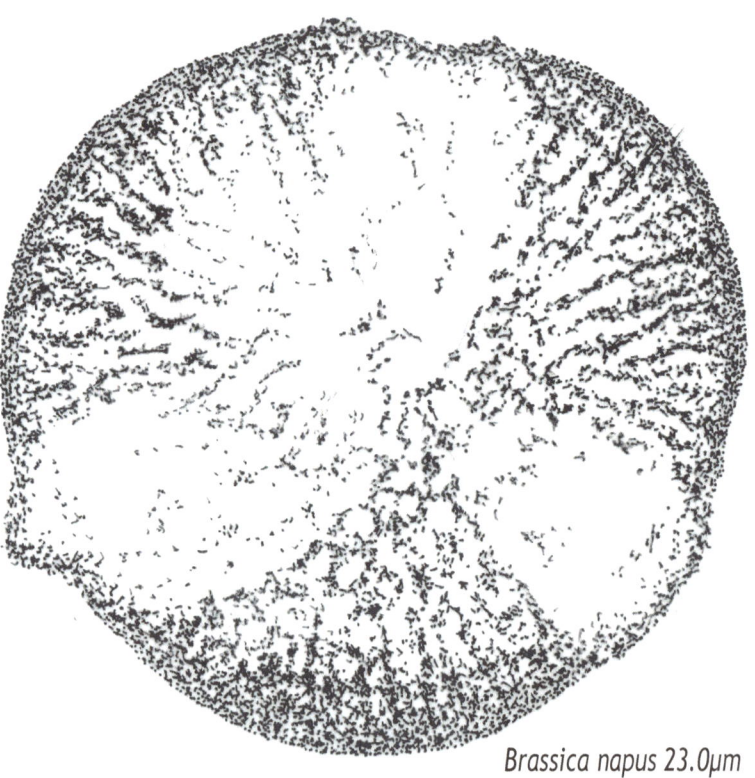

Brassica napus 23.0μm
Brassicaceae (Cruciferae)

Distinctive fluorescent yellow fields of brassica flowering in spring. Bees foraging on oil seed rape produce a pale, crystalline honey with rather a bland flavour.

Old man's beard

Untidy, rambling, hardy, the only clematis native to the UK. The pollen grain has a slightly knobbly surface which continues across the three indented furrows.

Clematis vitalba 20.7µm
Ranunculaceae

Pink oxalis

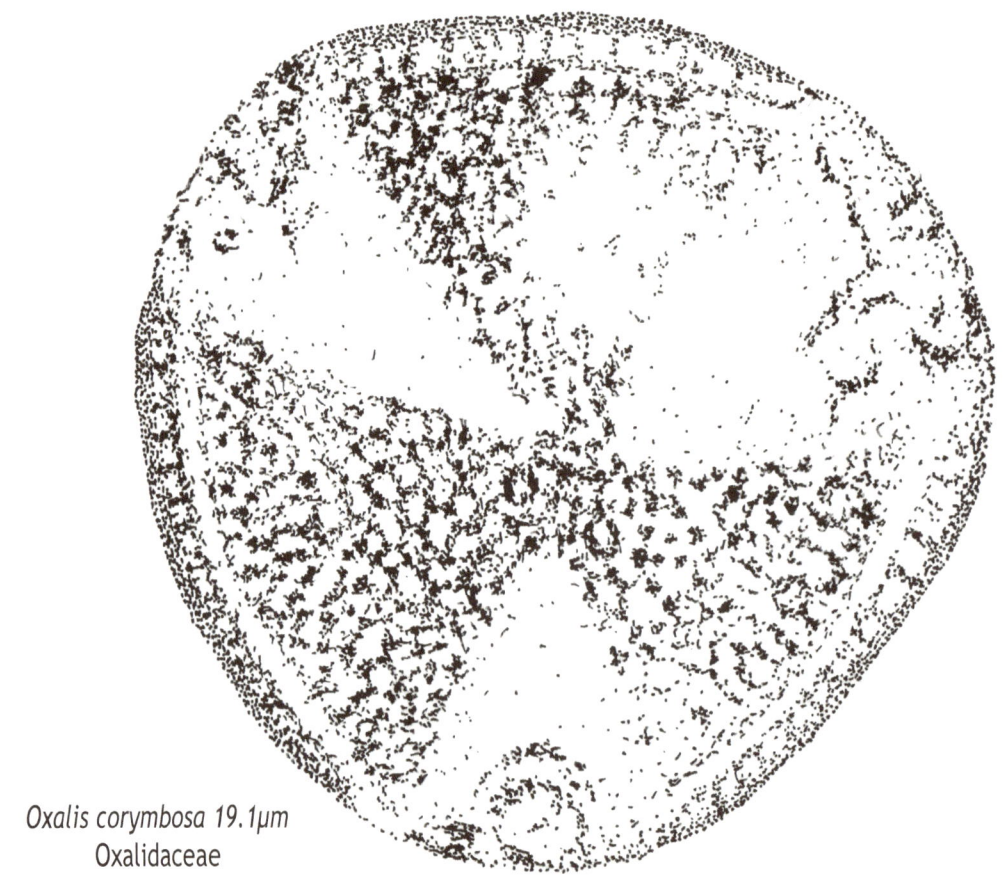

Oxalis corymbosa 19.1µm
Oxalidaceae

Long flowering member of the wood
sorrel family, which can
become invasive.
The pollen grain surface is netted
and the three furrows have a distinct
pattern of their own.

Passion Flower

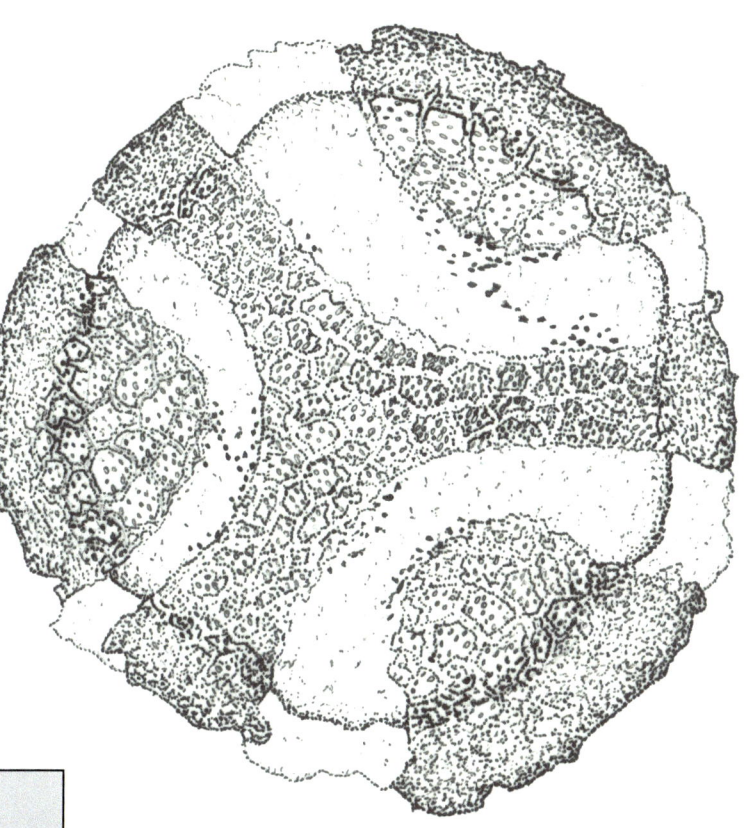

Passiflora caerulea 80.0μm
Passifloraceae

Photos © Sally Dunn

Neither native to, nor fully hardy in the UK,
the passion vine produces spectacular flowers
and edible fruits in warm summers. The
pollen grains have a unique curving pattern
resembling a tennis ball,
with three germinal furrows running between
the detailed sculpture of the outer wall
(exine).

Penstemon

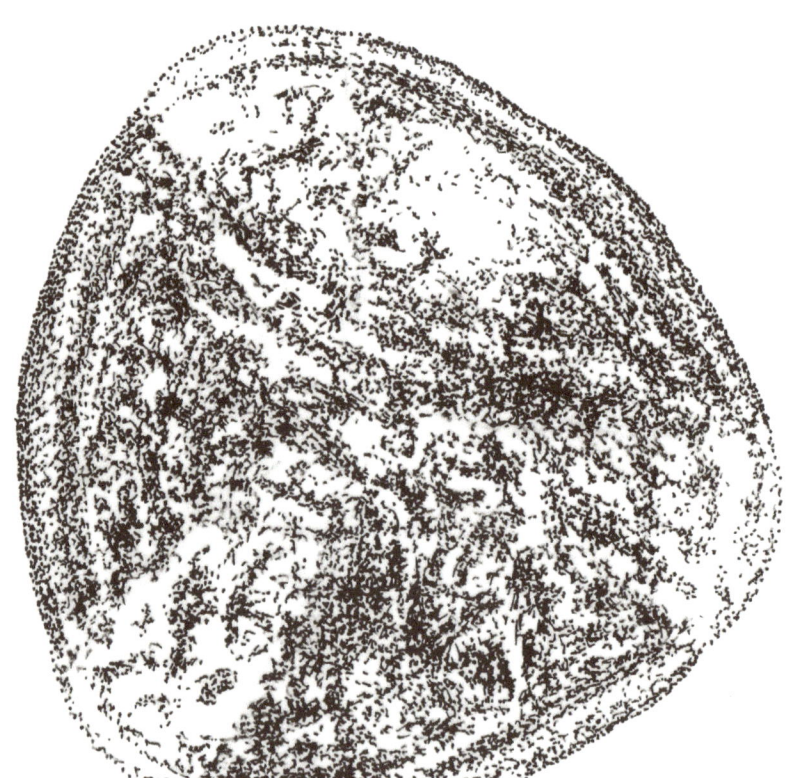

Penstemon sp. *19.6µm*
Plantaginaceae

Small triangular pollen grain with a common formation of furrows at the corners.

Photos © Sally Dunn

Periwinkle

Popular ground cover plant. The large pollen grain shows a polar view with a hint of the three furrows round the edge which would be along the equatorial axis.

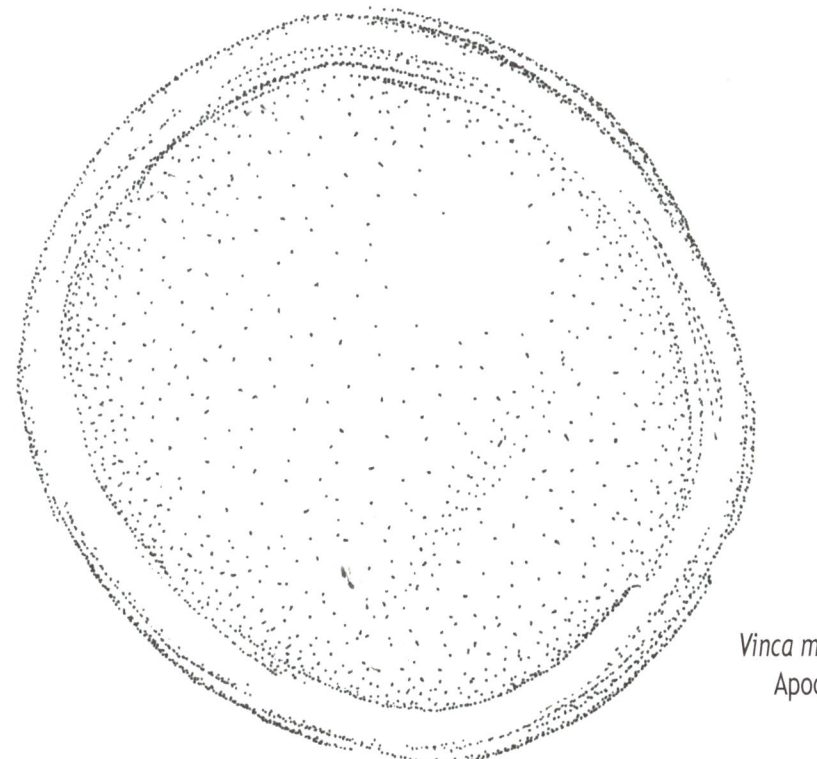

Vinca minor 96.0μm
Apocynaceae

Philadelphus (Mock orange)

Large shrub with heavily scented flowers. Quite small pollen grain roughly triangular in the polar view shown with a reticulate pattern over the surface.

Philadephus coronarius 16.0μm
Hydrangeaceae

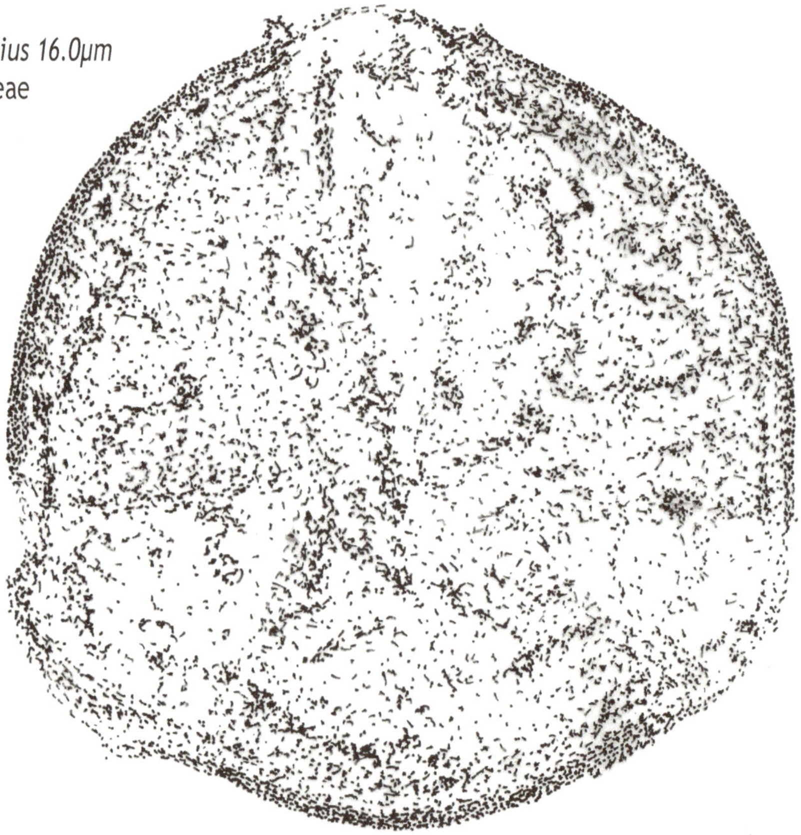

Pieris japonica

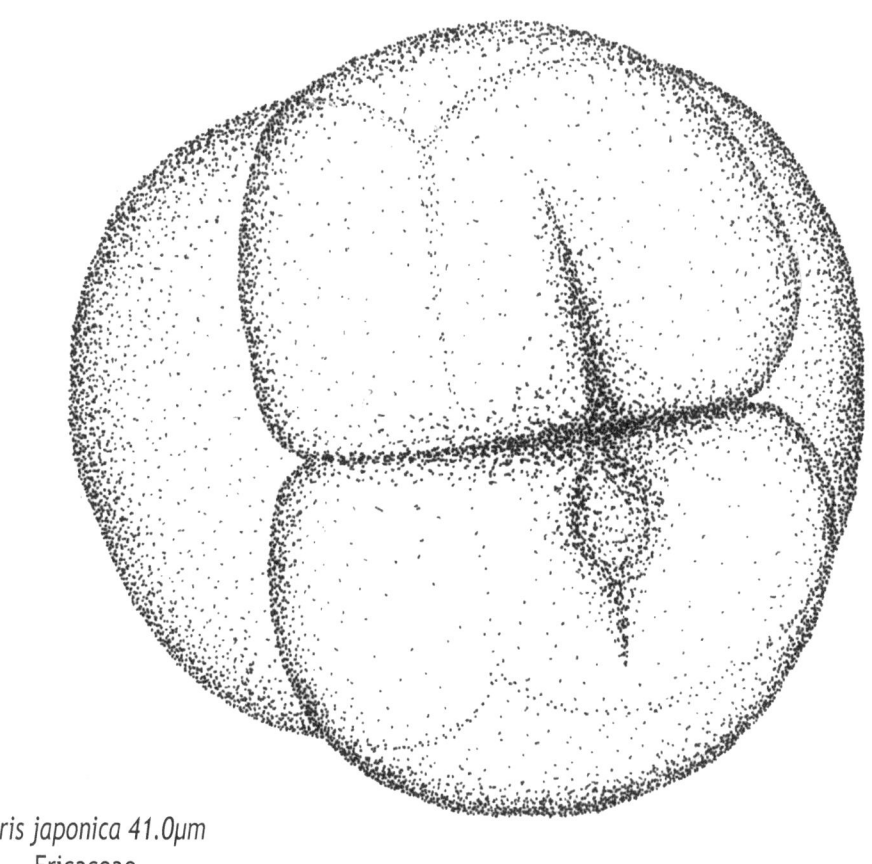

Pieris japonica 41.0μm
Ericaceae

An evergreen shrub, often with striking pink or red new leaf growth, thriving in acid soil.
It has the typical polyad pollen formation of the Erica family.

London Plane

Monoecious:
the yellow male flowers (left) grow from older branches further back toward the trunk and red female flowers (right) grow from newer shoots.
Polar view of the pollen grain showing the three furrows and the reticulated surface.

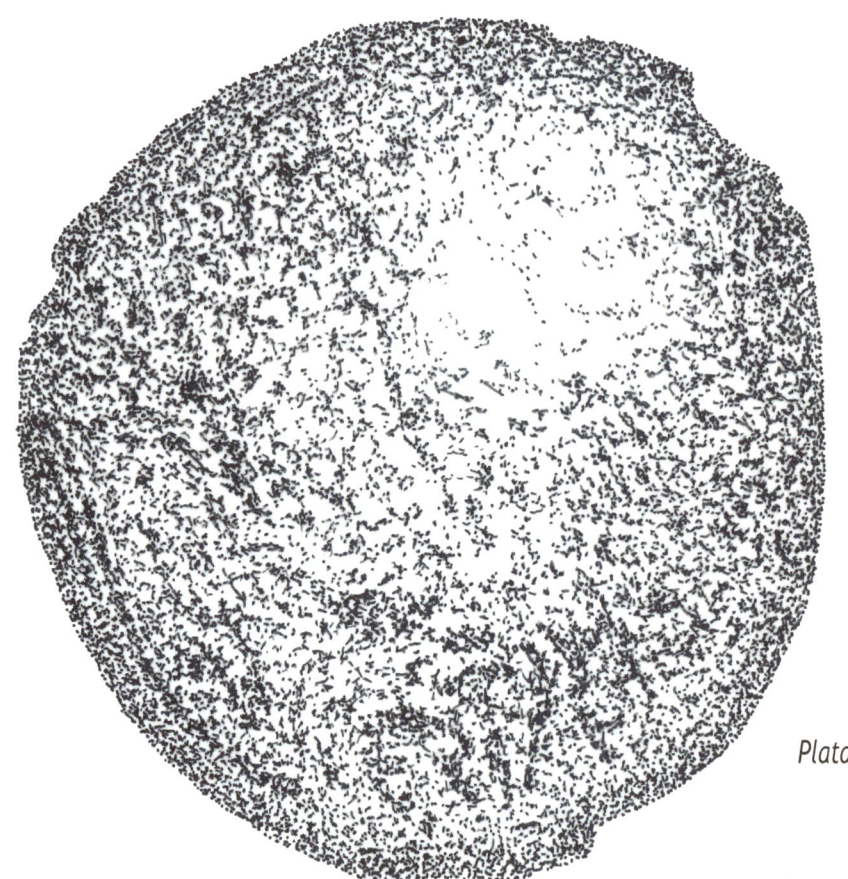

Platanus x hybrida 23.8µm
Platanaceae

Plum

Along with many other fruit trees, plums rely on undamaged (eg by frost) blossom, another, suitable, pollinator tree nearby, and insect vectors, in order to set fruit.
The reticulated surface of the grain has a striated pattern, and the furrows a surface webbing either side of the pore.

Prunus domestica 45.2μm
Rosaceae

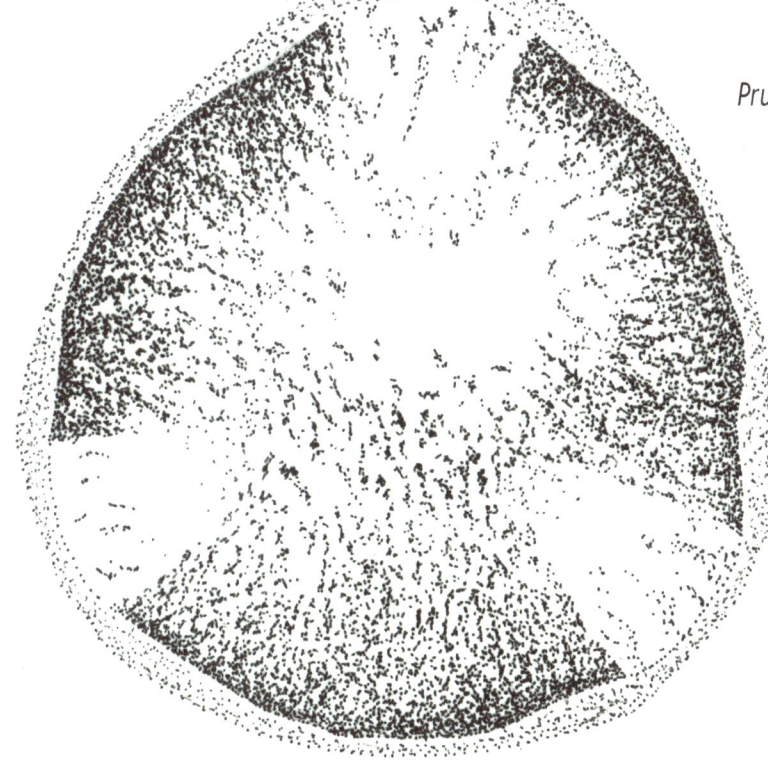

Poppy

Poppies have no nectaries but produce large quantities of pollen in shades of dark blue to grey. Pollen pellets collected from hives where bees have been foraging on poppies are often quite black.
The pollen grain is spheroid, reticulate and has three furrows.

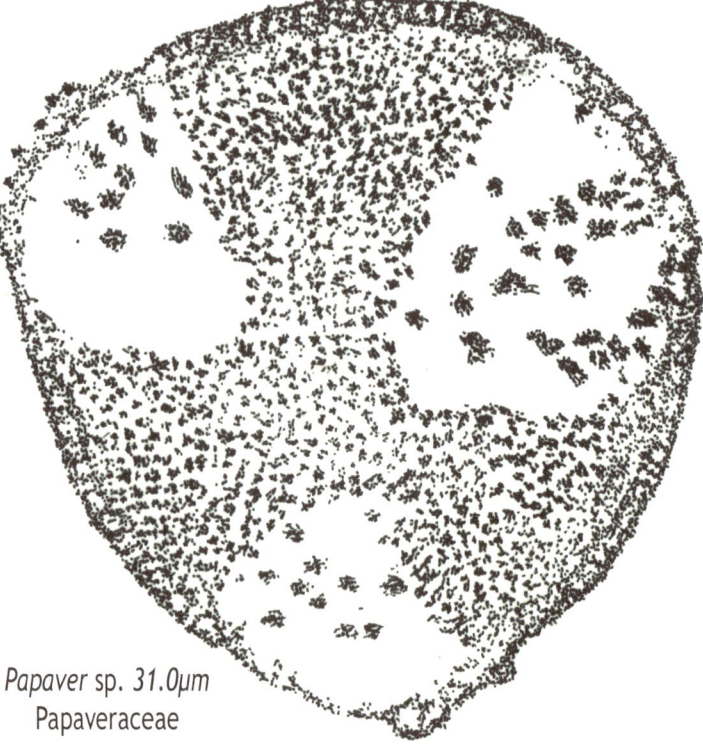

Papaver sp. 31.0μm
Papaveraceae

Privet

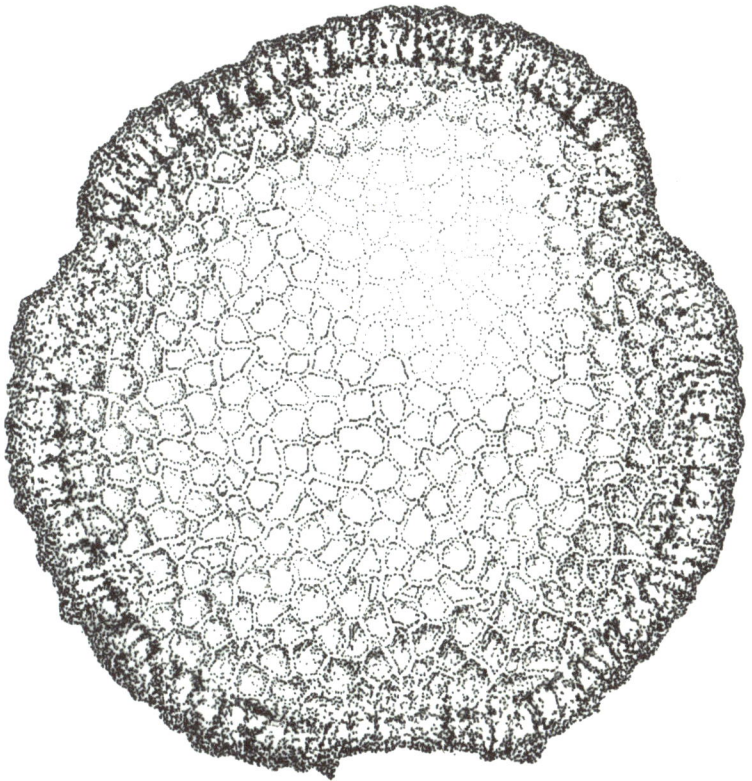

Ligustrum vulgare 27.0μm
Oleaceae

Popular hedge plant, usually trimmed before it can flower.
Note the net like surface of the pollen grain.

Pulmonaria

Pulmonaria angustifolia 37.7μm
Boraginaceae

A vigorous, clump forming semi-evergreen flowering perennial, also known as lungwort. The leaves are often spotted. The pollen grain shape is prolate or elongate so lies on its side with the view under the microscope always an equatorial view.
Around the equator, the grain has a slight ridge with unique surface pattern and four pores equally spaced around the equator.

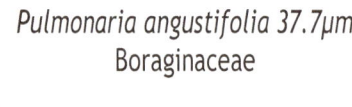

Pyracantha

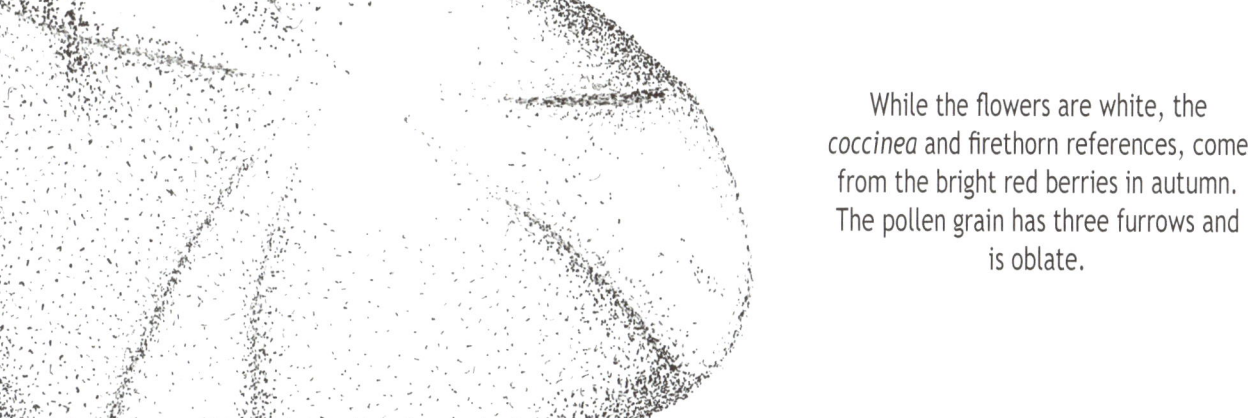

While the flowers are white, the *coccinea* and firethorn references, come from the bright red berries in autumn. The pollen grain has three furrows and is oblate.

Pyracantha coccinea 24.7µm
Rosaceae

Ragwort

Senecio jacobaea (Jacobaea vulgaris) 27.4µm
Asteraceae (Compositae)

A distinctive, common, poisonous, persistant, roadside and meadow weed, unpalatable to most animals. If eaten in quantity, liver damage accumulates but it is seldom a problem unless contaminating hay. The pollen grain appearance is typical of the daisy family.

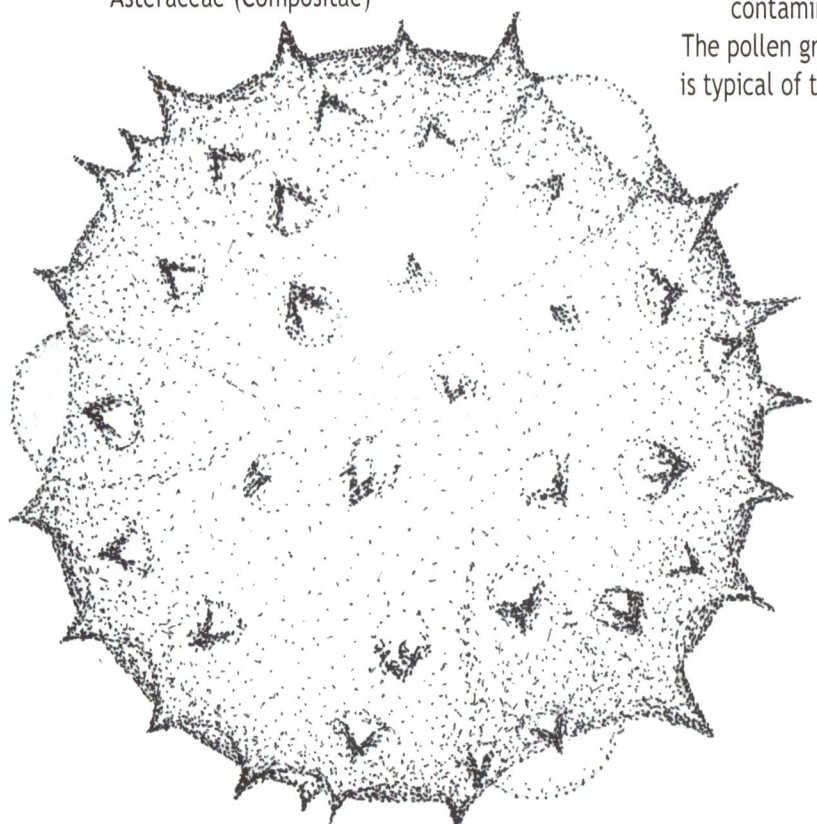

Red currant

Deciduous bush about 1.5m high,
producing flowers in springtime
attractive to insects,
followed by racemes of edible red
currants in early summer.
The pollen grains have
eight pores.

Ribes rubrum 33.5μm
Grossulariaceae

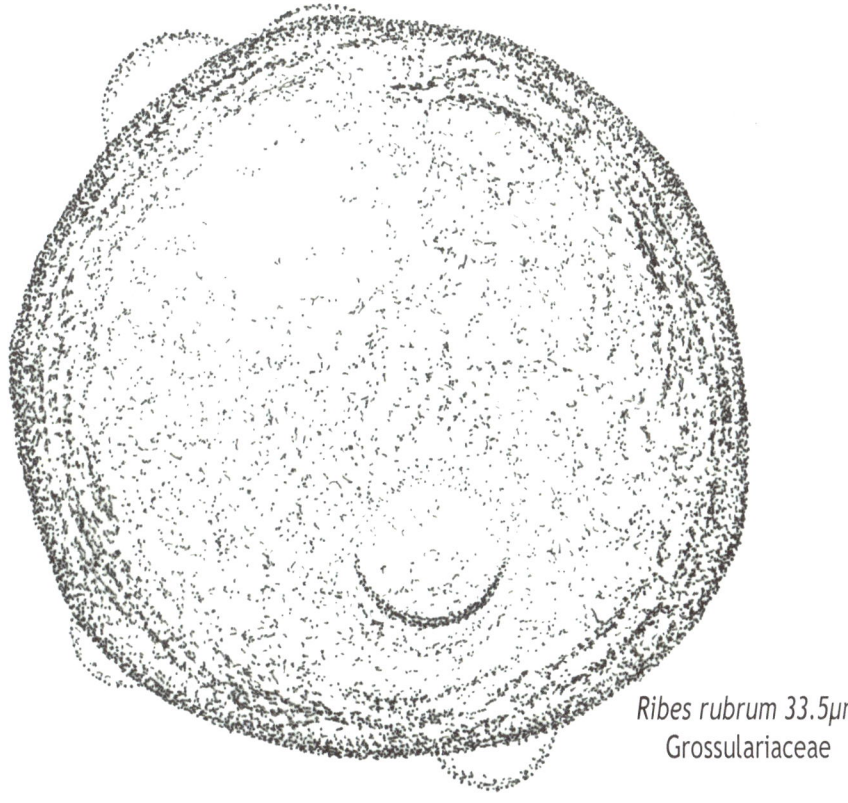

Redshank

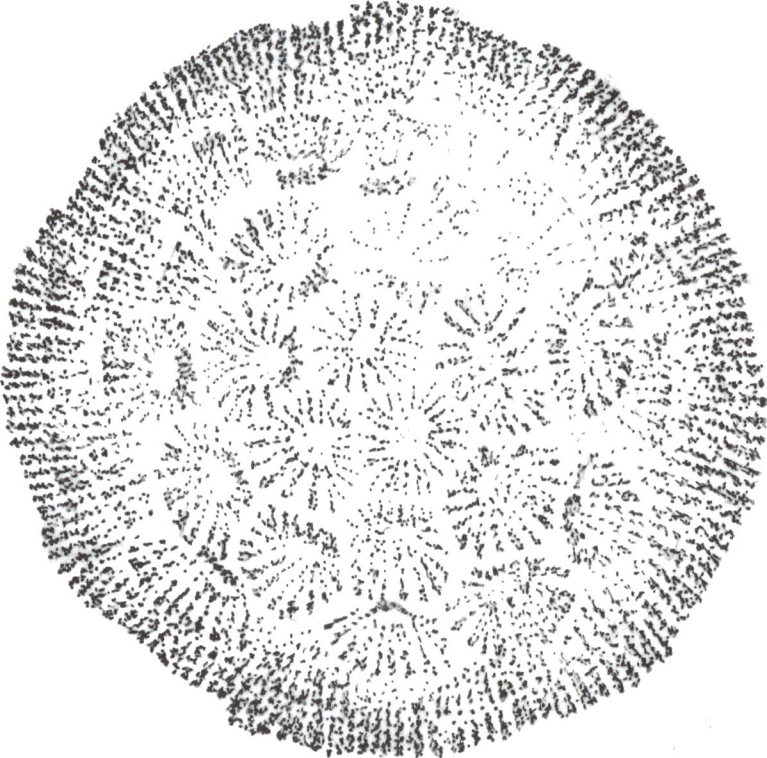

Persicaria maculosa (syn. Polygonum persicaria) 36.2 µm
Polygonaceae

A weed of cultivation, the tiny individual redshank flowers are not easy to dissect for their pollen.
The pollen grain surface has a regular reticulate pattern with some of the whorls containing pores.

Ribwort plantain

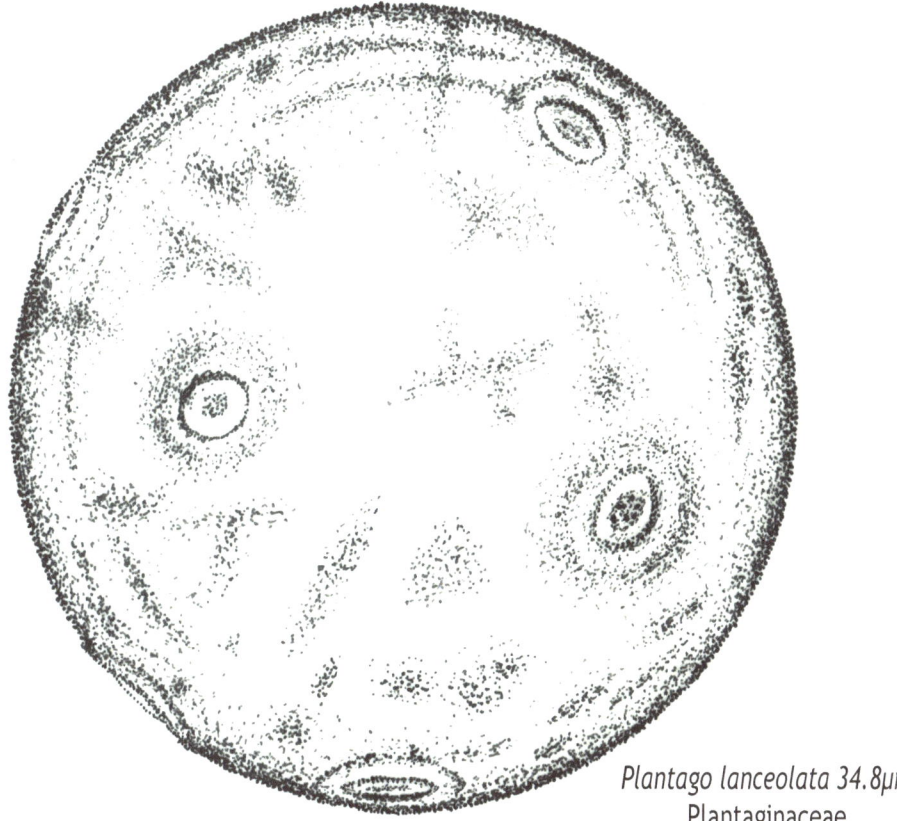

A perennial herb providing nutritious grazing for livestock and wildlife. The flower head passes for grass until examined more closely. The anthers form a protruding whorl around the flower stem. The pollen grain has nine pores.

Plantago lanceolata 34.8µm
Plantaginaceae

Rhododendron

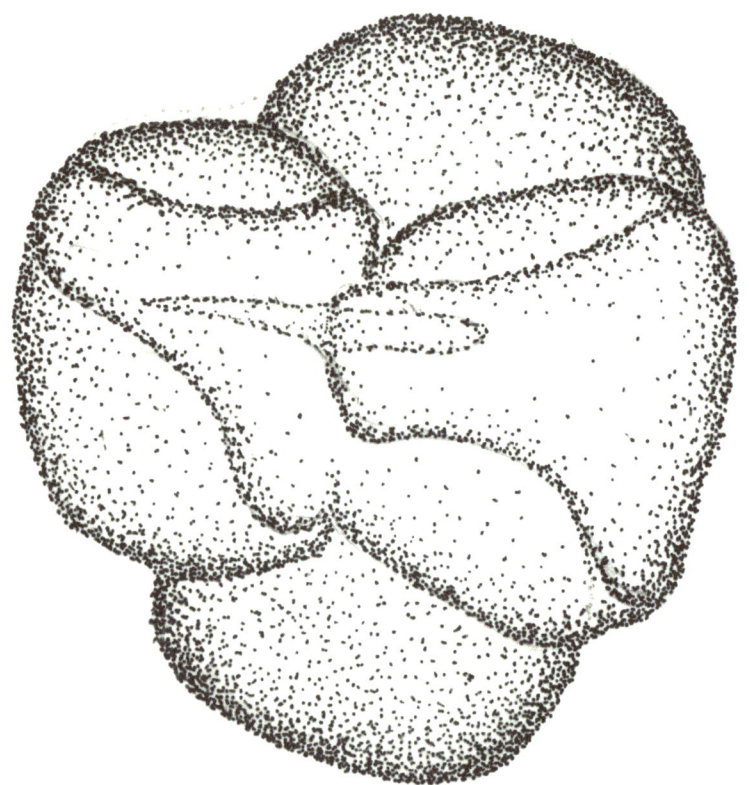

Rhododendron ferrugineum 62.0µm
Ericaceae

Woody bushes producing heavily scented, spectacular flowers in late spring.
Tetrad pollen formation, formed from anthers in sets of four, typical of the Ericaceae.

Ramanus rose

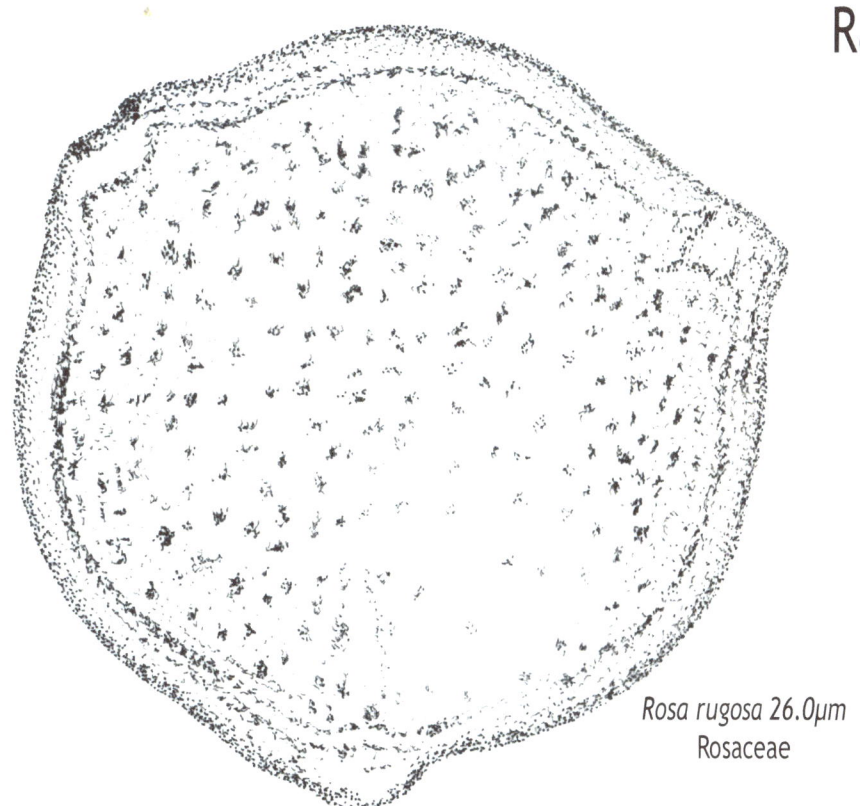

Rosa rugosa 26.0μm
Rosaceae

Dense, prickly bush sometimes used for hedging, attractive to bees and butterflies.

Rock rose

Drought tolerant long flowering plant. Equatorial view of the pollen grain showing the pore at the centre of the furrow and the netted reticular pattern of the surface of the grain.

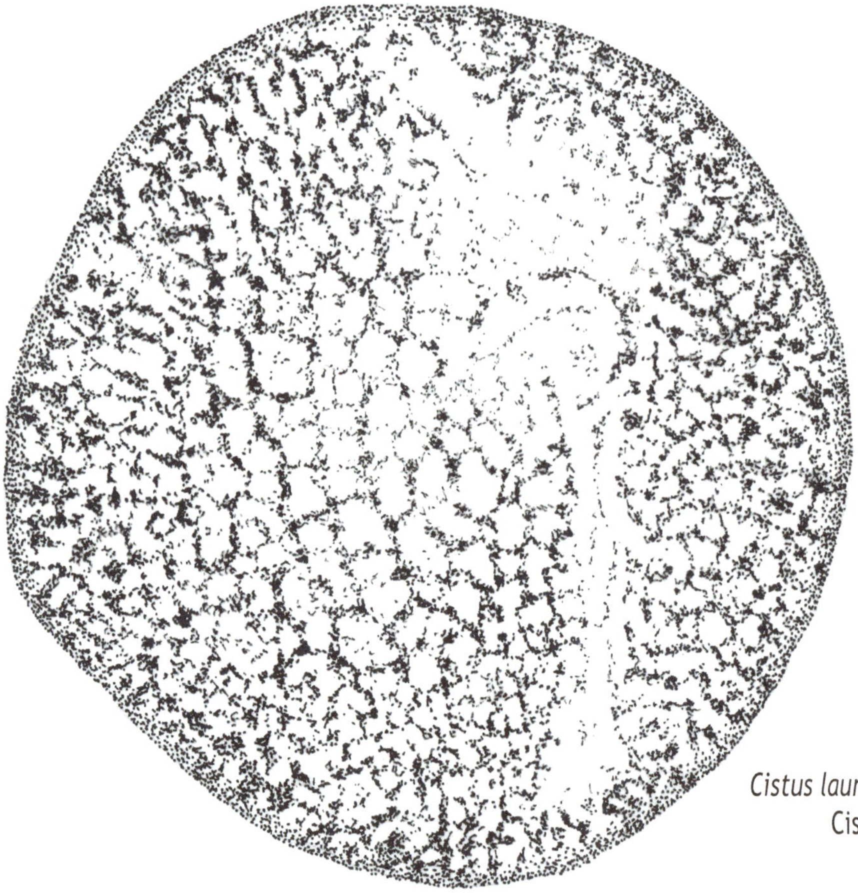

Cistus laurifolius 48.7μm
Cistaceae

Rosemary

Well known, almost hardy, shrubby culinary herb. The pollen grain has six furrows.

Rosmarinus officinalis 43.0μm
Lamiaceae

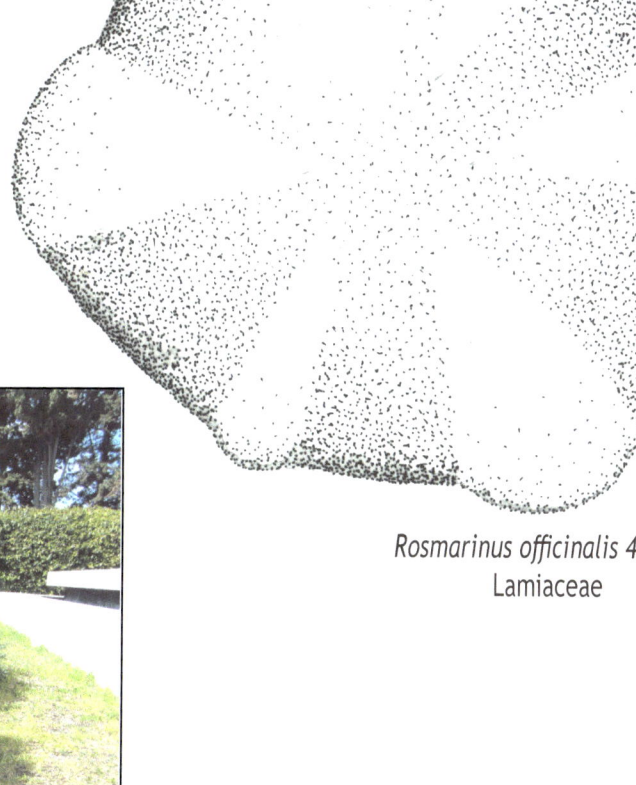

Sainfoin

Once grown as a forage crop, the common sainfoin is an open pollinating plant, ie produces offspring just like the parent when self pollinated or pollinated from a similar plant. It is mainly pollinated by nectar feeding insects.
The pollen grain is eliptical in equatorial view. The three longitudinal furrows have a line pattern down their centre.

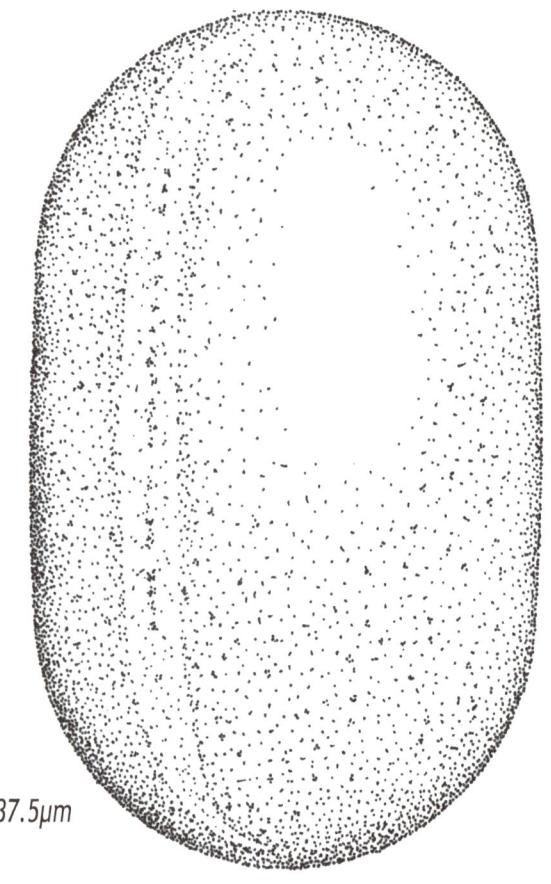

Onobrychis viciifolia 37.5µm
Fabaceae

Field scabious

Knautia arvensis 12.2µm
Caprifoliaceae

Perennial plant found in grassy areas, preferring dry soils, the pollen grain is smaller than average.
It has three pores with tufts growing from the centre of each.

Scarlet pimpernel

A low growing annual, the pollen grain has a reticulated surface and three furrows with pores at the equator.

Anagallis arvensis 20.8μm
Primulaceae

Scentless mayweed

The few spikes of this pollen grain are arranged in regular patterns around the three pores.

Tripleurospermum inodorum (Matricaria perforata) 29.0µm
Asteraceae (Compositae)

Photo © Sally Dunn

Scots pine

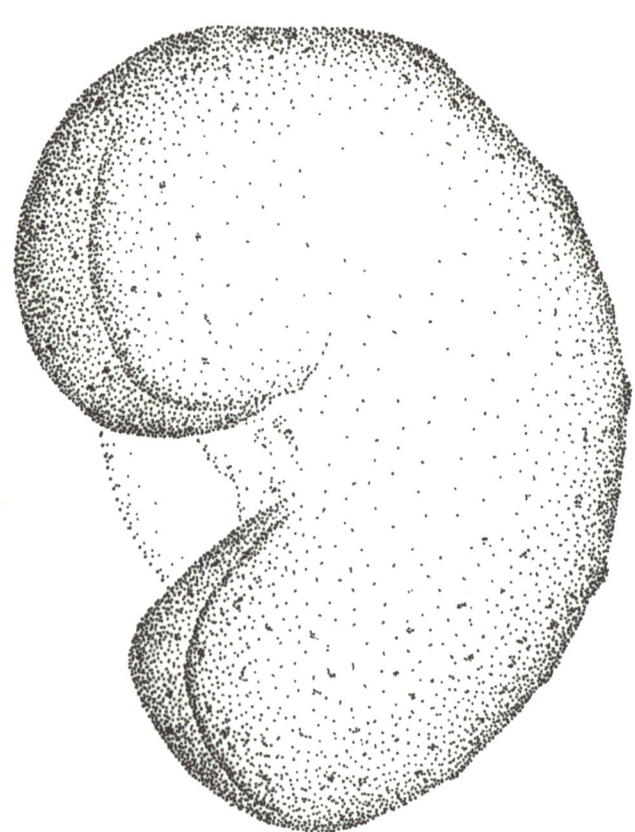

Pinus sylvestris 88.0µm
Pinaceae

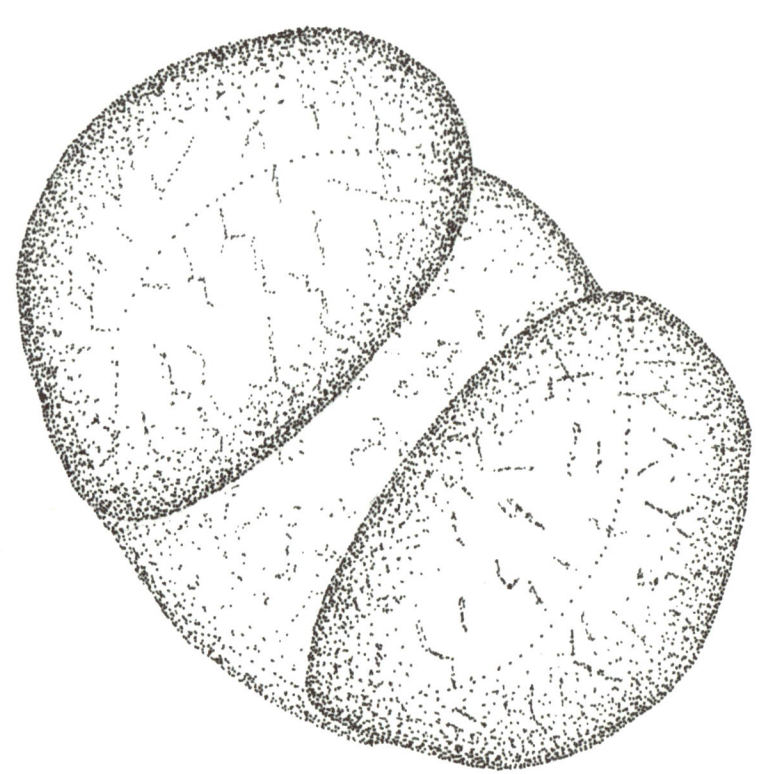

Pine pollen has air-filled bladders that help its dispersal in the wind but also allow it to float on water and wash up on to river banks. Distinguishable by size and shape, this pollen often turns up as a stray among other pollens and out of its flowering time.

Silver birch

Betula pendula 28.5µm
Betulaceae

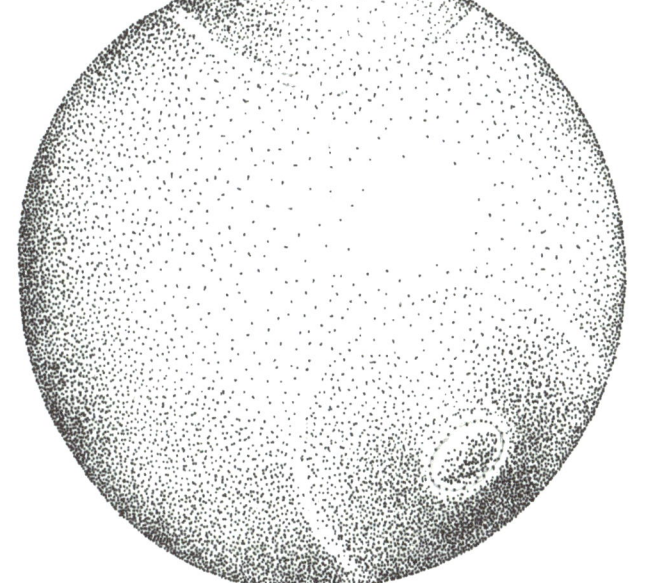

A medium sized deciduous tree with distinctive white, peeling bark. Separate male and female catkins are produced on the same tree. Male catkins form in the autumn, ripening and opening to release the pollen in the spring. Birch pollen is well known as a common allergen. Polar view top left, equatorial view above showing two of the three pores of the pollen grain.

Skullcap

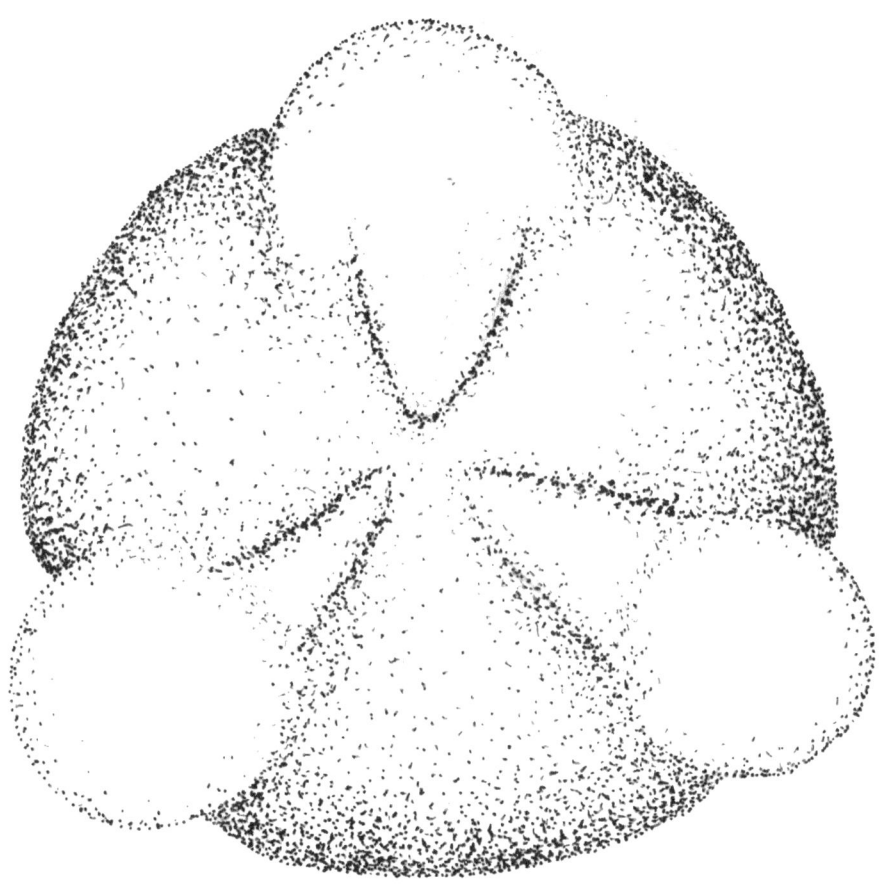

Scutellaria galericulata 24.6 μm
Lamiaceae

A member of the mint family with quite long, up to 2cm, blue flowers. Unlike some members of the mint family, skullcap pollen has only three furrows.

Snake's head fritillary

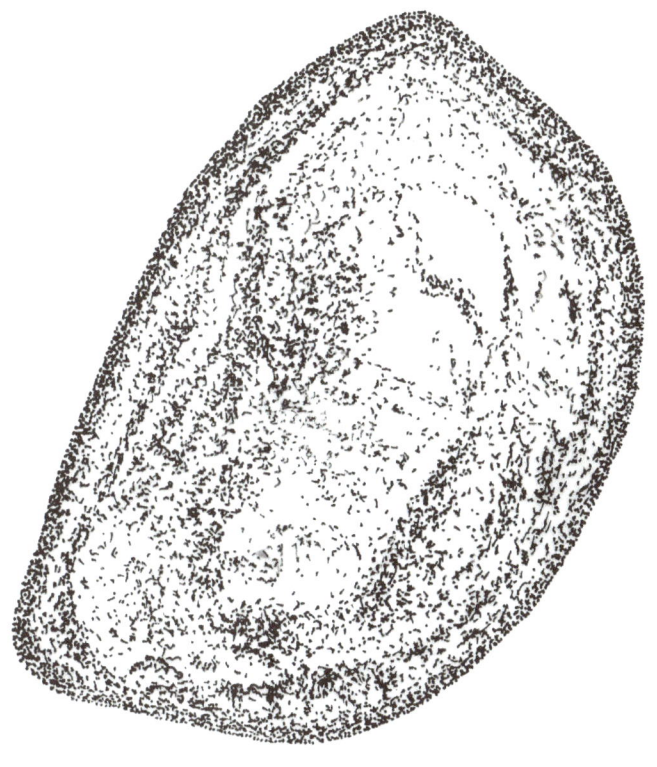

Fritillaria meleagris equatorial view above 48.6µm; distal polar view below 47.9µm
Liliaceae

The pollen grain is boat shaped with one wide sunken aperture taking up about a fifth of the surface stretched across one pole; and a lattice structure over the surface.

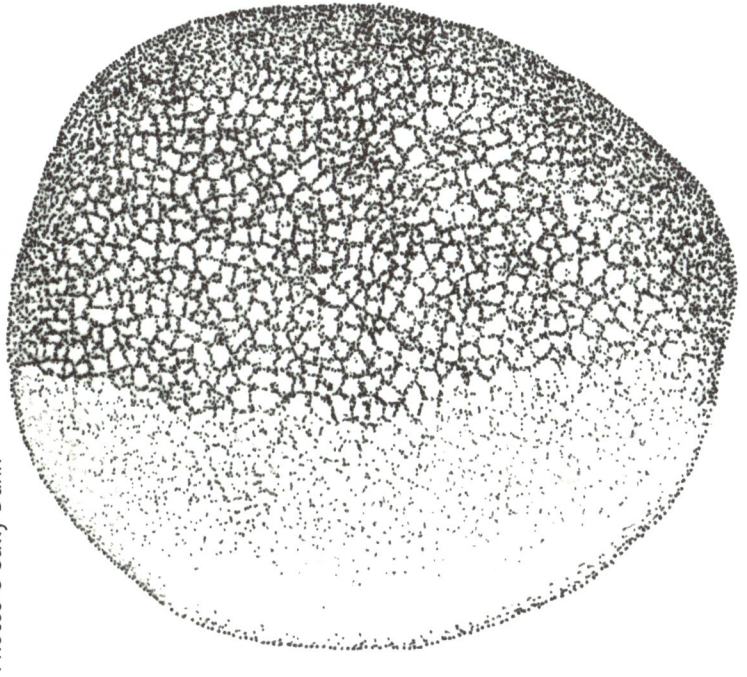

Photos © Sally Dunn

Snapdragon

The snapdragon depends mostly on bumble bees for its fertilisation. The stronger scent in daytime, and the colour (most often, naturally, yellow) of the flower attract bumble bees. The design of the flower requires the weight of the bumble bee to press down and open it, and the shape allows the bee to reach in for the nectar, while rubbing pollen onto itself to take to the next flower.

Antirrhinum majus 22.0µm
Plantaginaceae

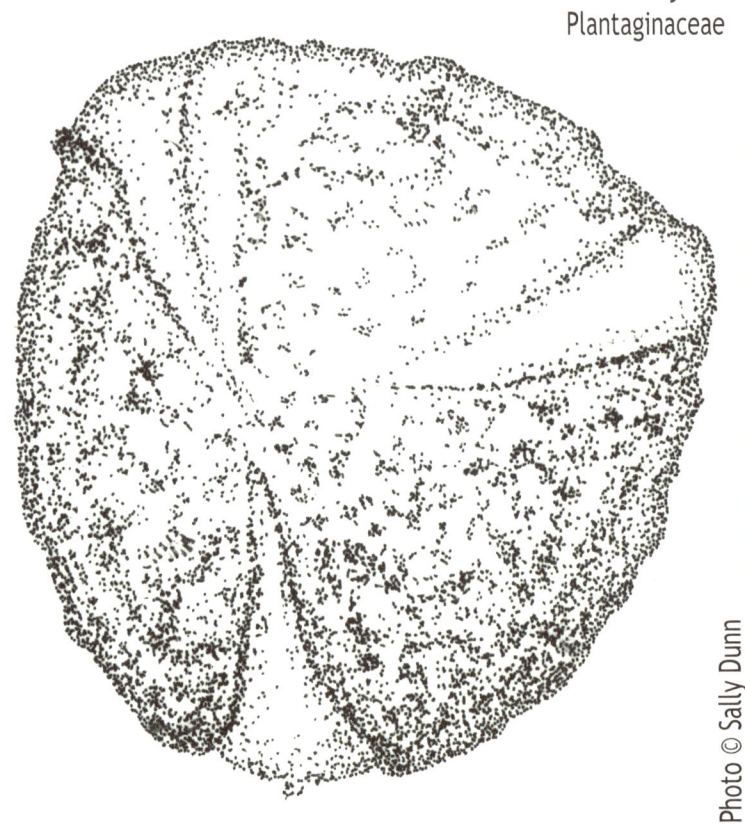

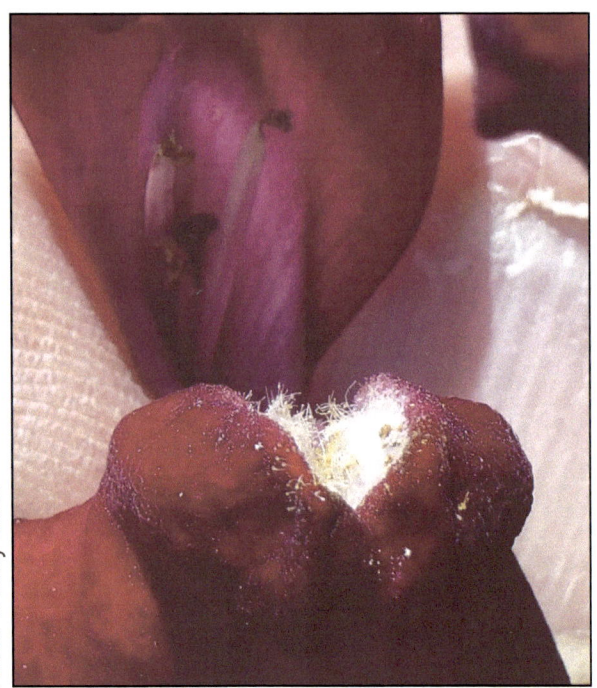

Photo © Sally Dunn

Snowdrop

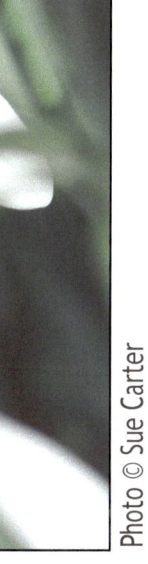

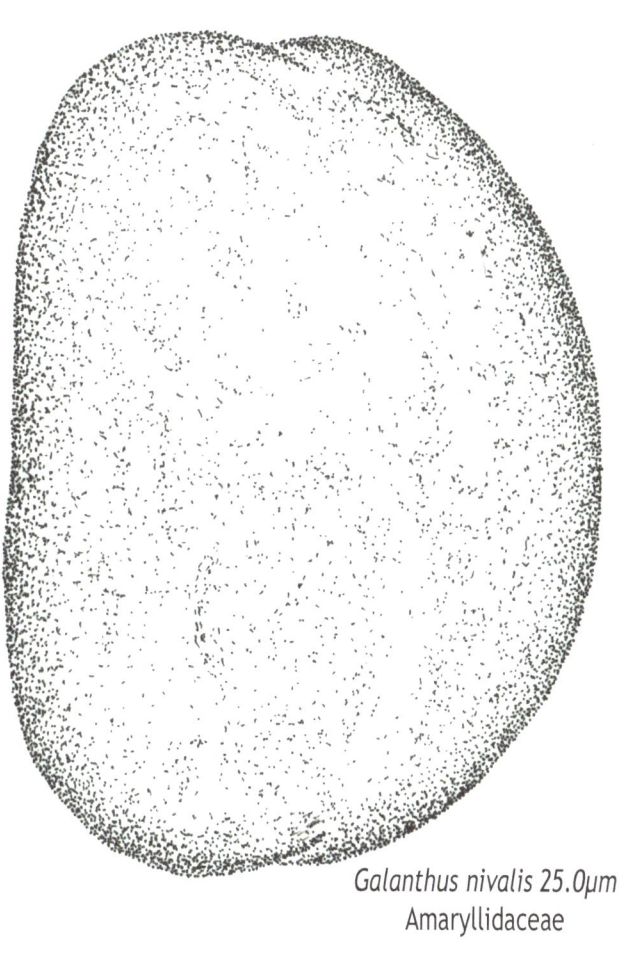

Galanthus nivalis 25.0µm
Amaryllidaceae

The pollen grain's single furrow is typical of the Amaryllidaceae. The pollen is an orange/red colour.

Snowy mespilus

A shrub or small tree. The pollen grain furrows are mostly taken up with the large protruding pores.

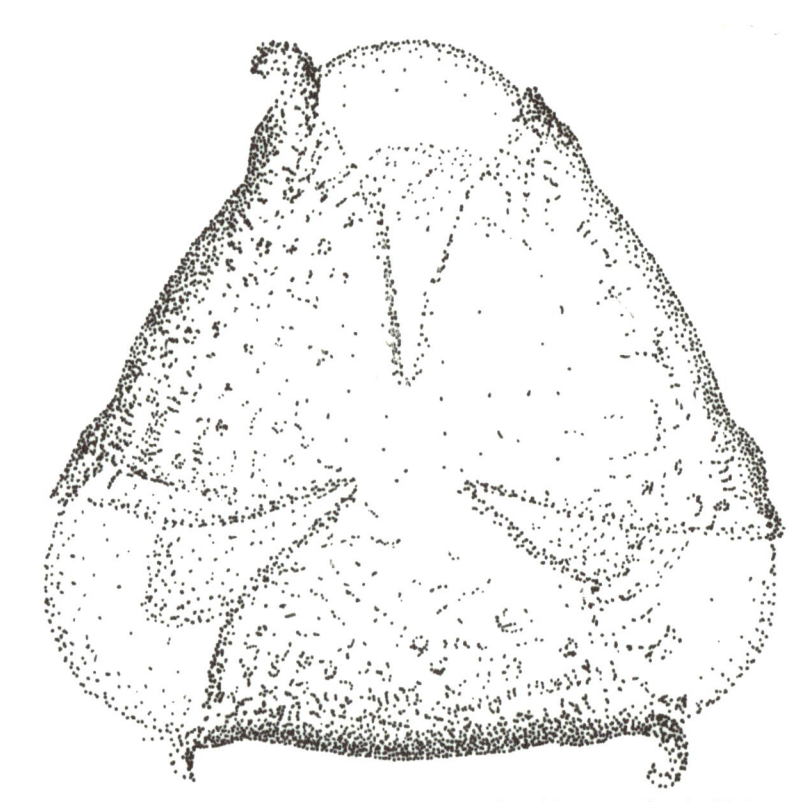

Amelanchier canadensis 39.5µm
Rosaceae

Photo © Sally Dunn

Ivy leaved speedwell (blue)

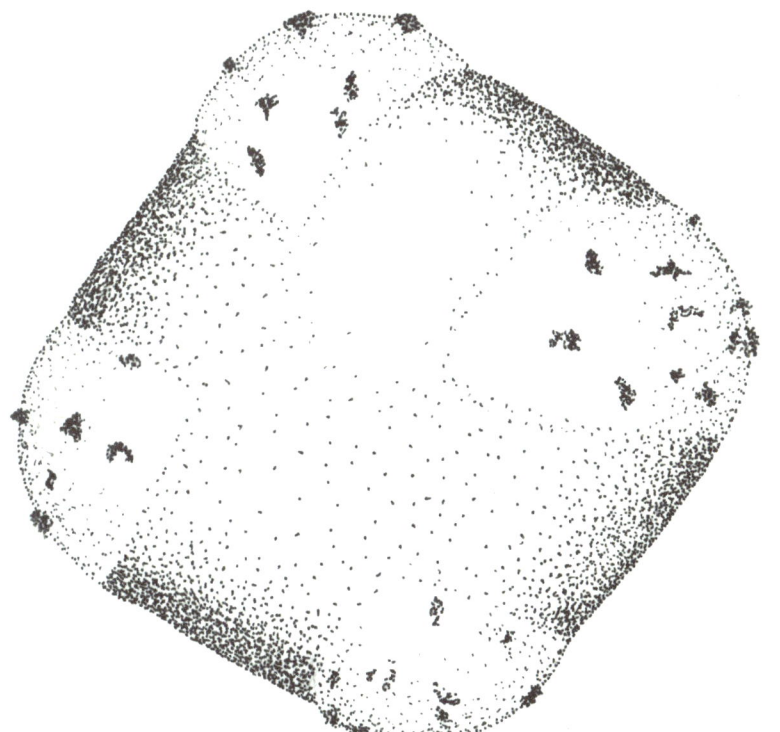

Veronica hederifolia 25.2µm
Plantaginaceae

A common weed which sprawls outwards, rooting where the stems touch the earth, and which seeds readliy, the seeds persisting for several years.
Each pollen grain varies in number of decorated furrows, usually 3 or 4, often irregularly arranged. This grain has four wide furrows, one at each corner.

Speedwell (germander)

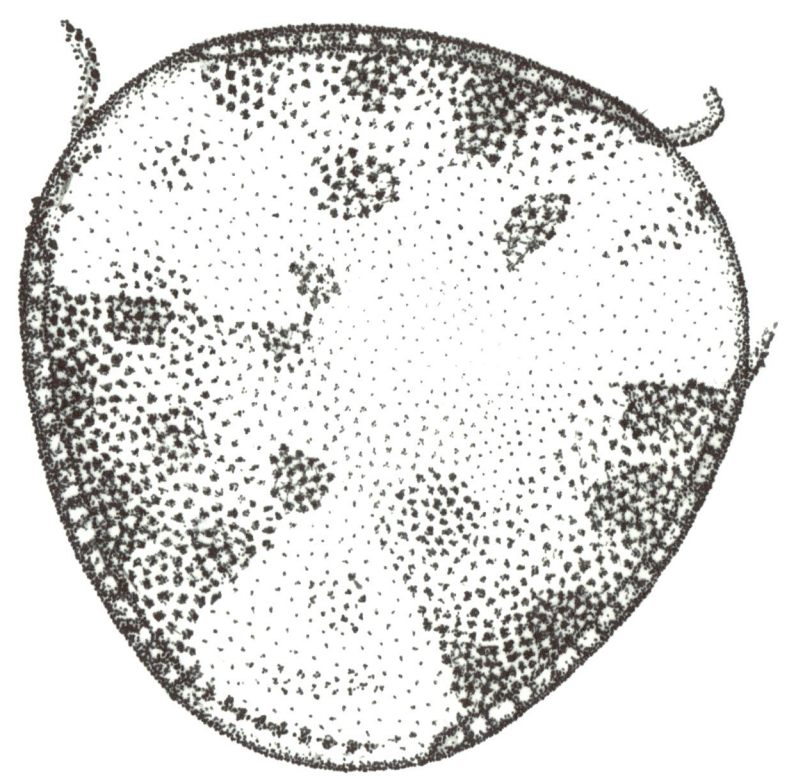

Usually low growing, perennial, invasive, often an unwelcome weed in lawns. Germander speedwell has flowers with all blue petals.
This pollen grain usually has three wide furrows, grains with four furrows are not uncommon.

Veronica chamaedrys 39.0μm
Plantaginaceae

Spindle tree

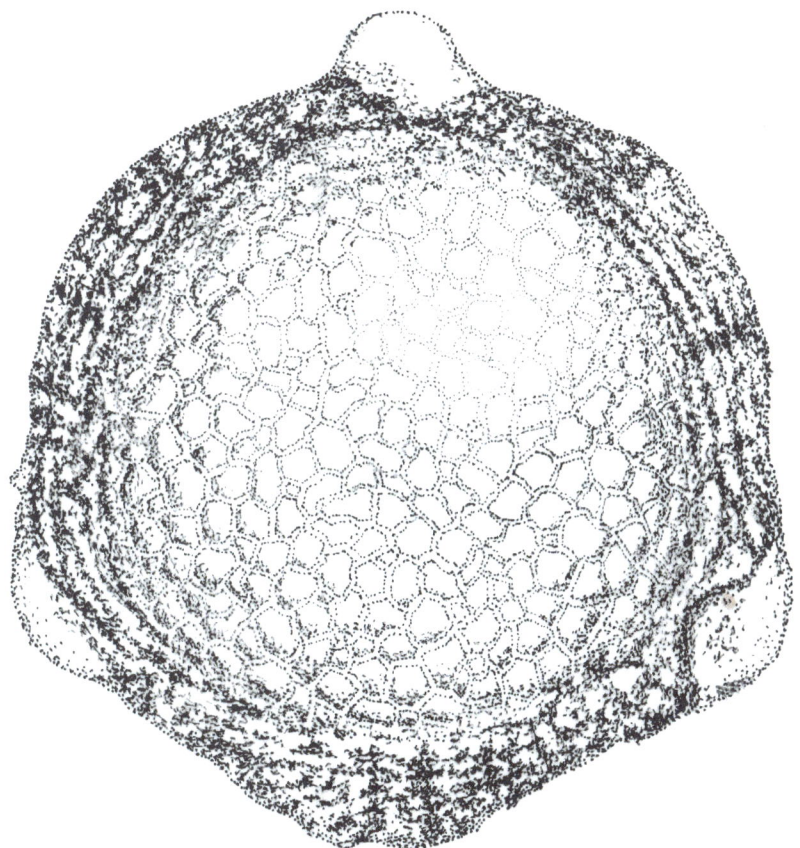

Euonymus europaeus 22.8μm
Celastraceae

Spindle trees are native to the UK, growing well on chalky soil. They are hermaphrodite, having flowers with both male and female parts. The small yellow-green flowers appear in May and June, growing in clusters. They have four petals and are pollinated by insects. The characteristic bright pink fruits with orange seeds help identification.
The pollen grain has a netted appearance.

Photo © Sally Dunn

Star of Bethlehem

The pollen grain is boat shaped with one sunken furrow.

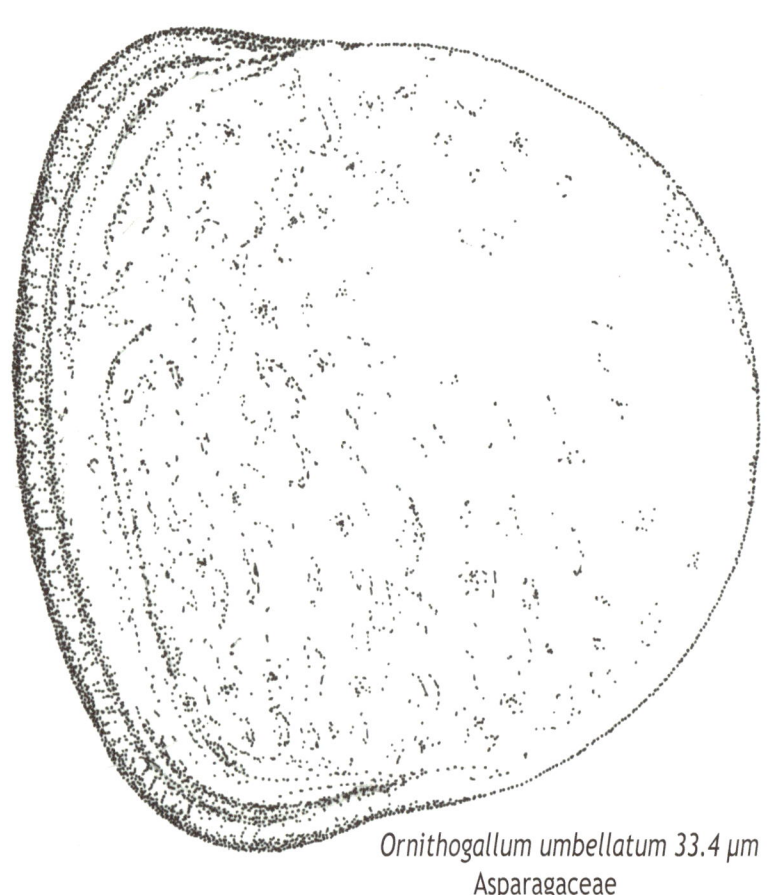

Ornithogallum umbellatum 33.4 µm
Asparagaceae

Strawberry

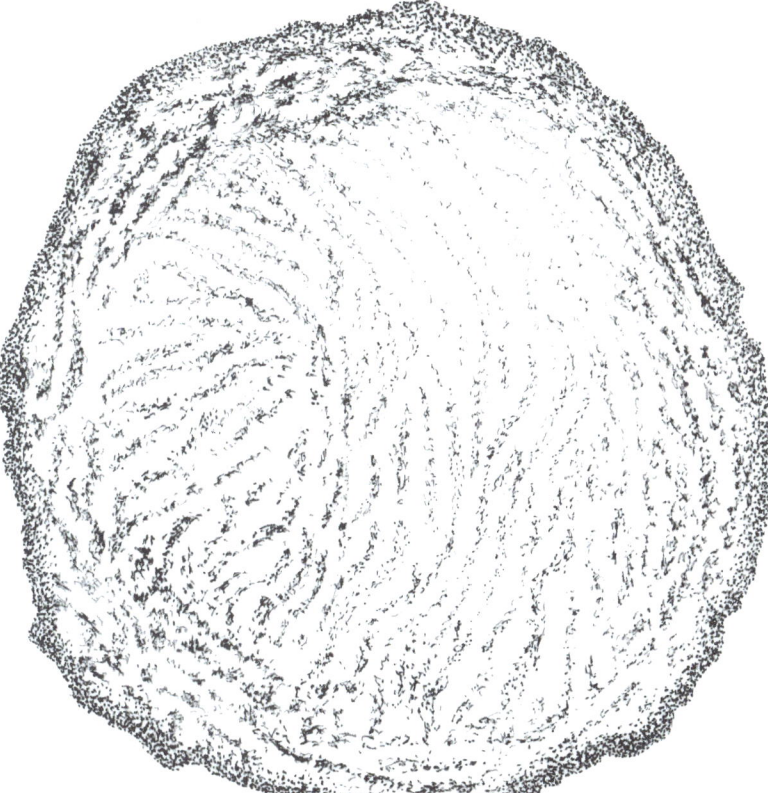

The formation of the much loved summer fruit of the strawberry depends upon insects visiting and pollinating the flowers. The more familiar white flowers have the same basic form as the pink variety illustrated here. The flower has multiple ovaries at its centre, each of which develops into a seed, of which there are many seen dotted around the outside of the fruit.
The pollen grain has a striated surface, the striations curve slightly around the not very obvious furrows, as seen in the one on the left of the drawing.

Fragaria × ananassa 22.3µm
Rosaceae

Strawberry tree

An evergreen shrub or small tree with small, bell shaped hermaphrodite flowers, pollinated by bees. The edible fruit looks like, but is not related to the common strawberry.
This tree has pollen with a very tightly packed tetrad formation with furrows aligned at midway points.

Arbutus sp. 46.0µm
Ericaceae

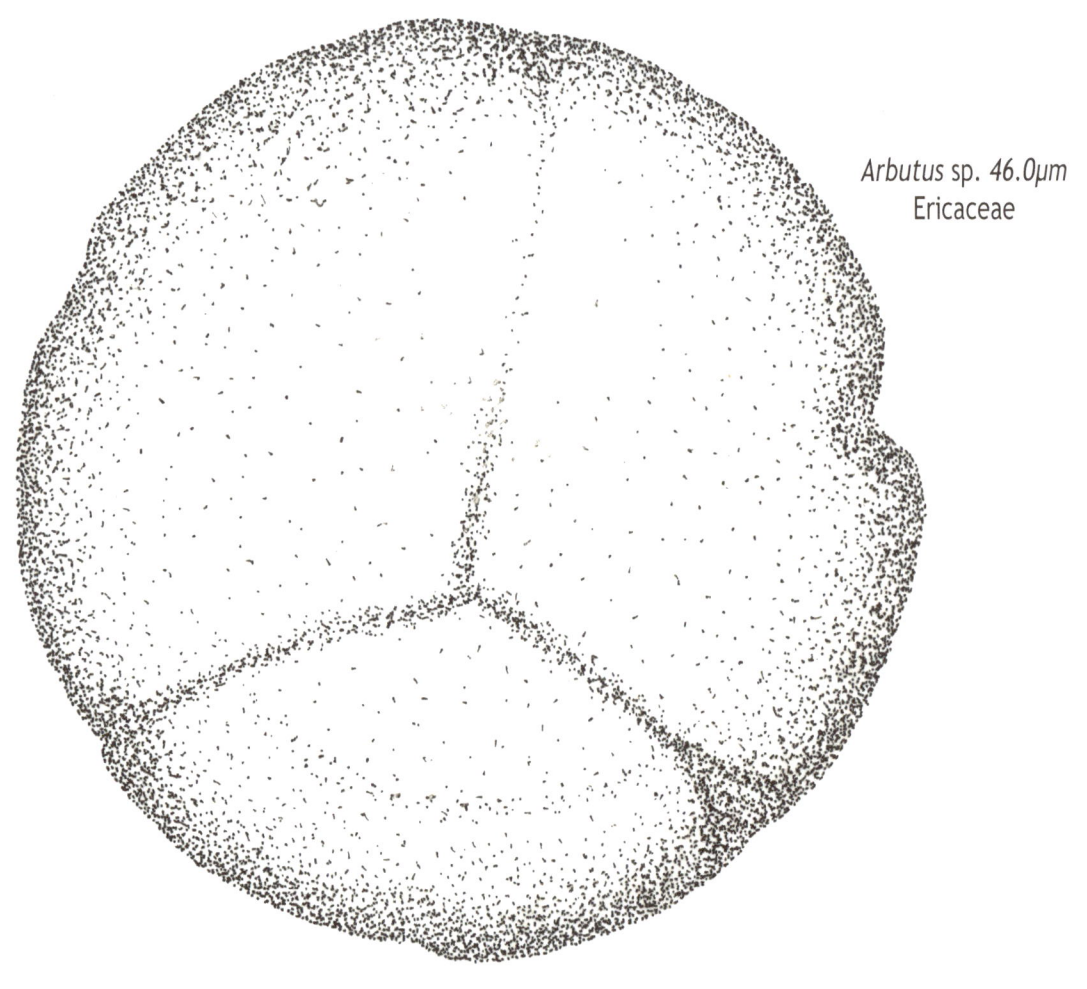

Stonecrop

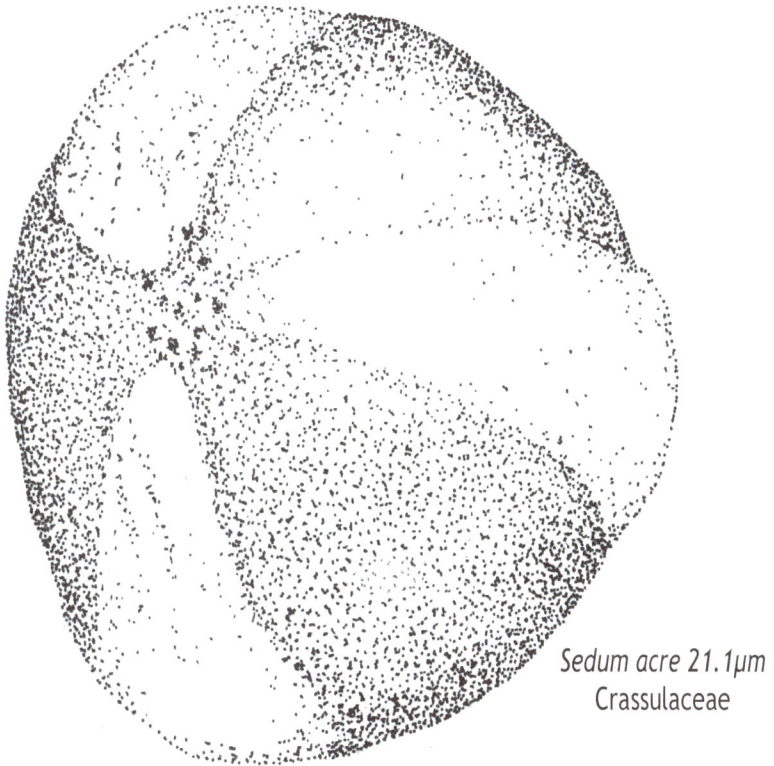

Sedum acre 21.1μm
Crassulaceae

Stonecrops are a large group of perennial succulents. The low, densely growing varieties are often used as ground cover.
The pollen grain of *Sedum acre* is small, having three quite wide furrows, each with a large pore at the equator and a criss-cross netting pattern over the rest of the surface.

Summer snowflake

Leucojum aestivum 30.8μm
Amaryllidaceae

A taller plant than the snowdrop, and flowering in late spring, the nodding, white, bell shaped flowers have green tips. The pollen grain has one furrow stretching from pole to pole.

Sunflower

Popular garden annual, also grown as a crop for its oil-rich seeds, the flowers provide good forage for bees. The pollen has the spiky exterior of the Asteraceae with three short furrows each having a large pore at its centre.

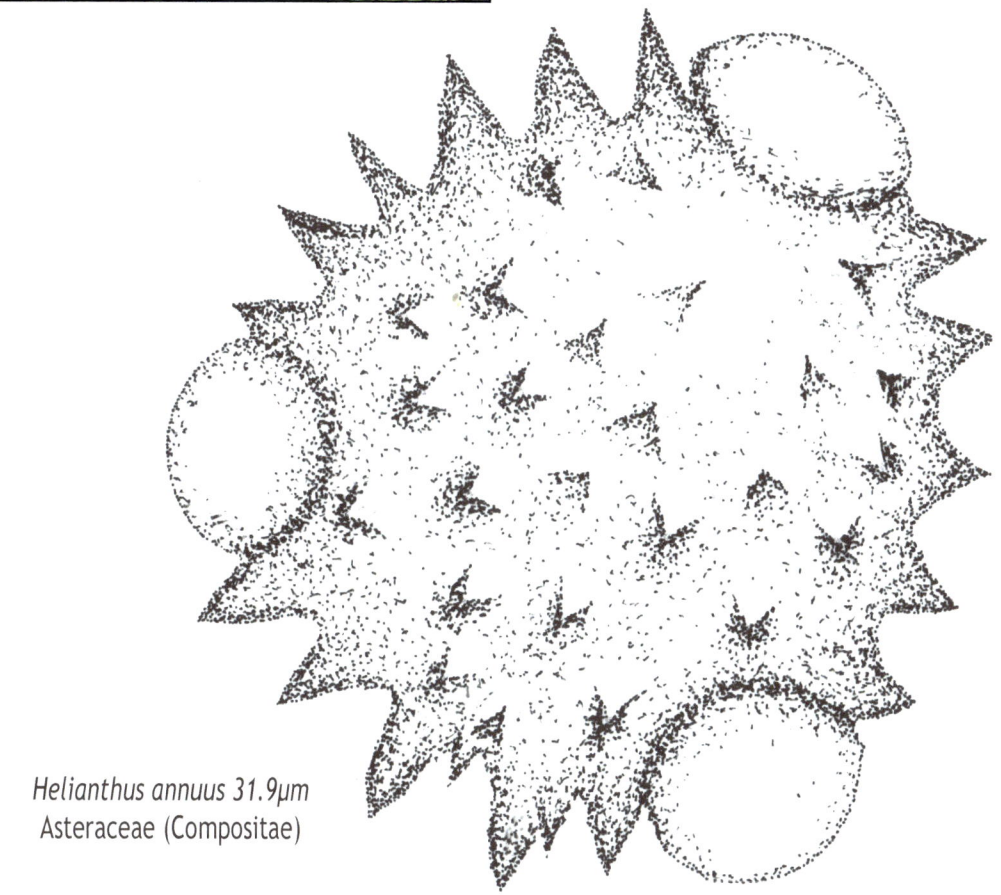

Helianthus annuus 31.9μm
Asteraceae (Compositae)

Sweet pea

Lathyrus odoratus 45.4μm
Papilionaceae

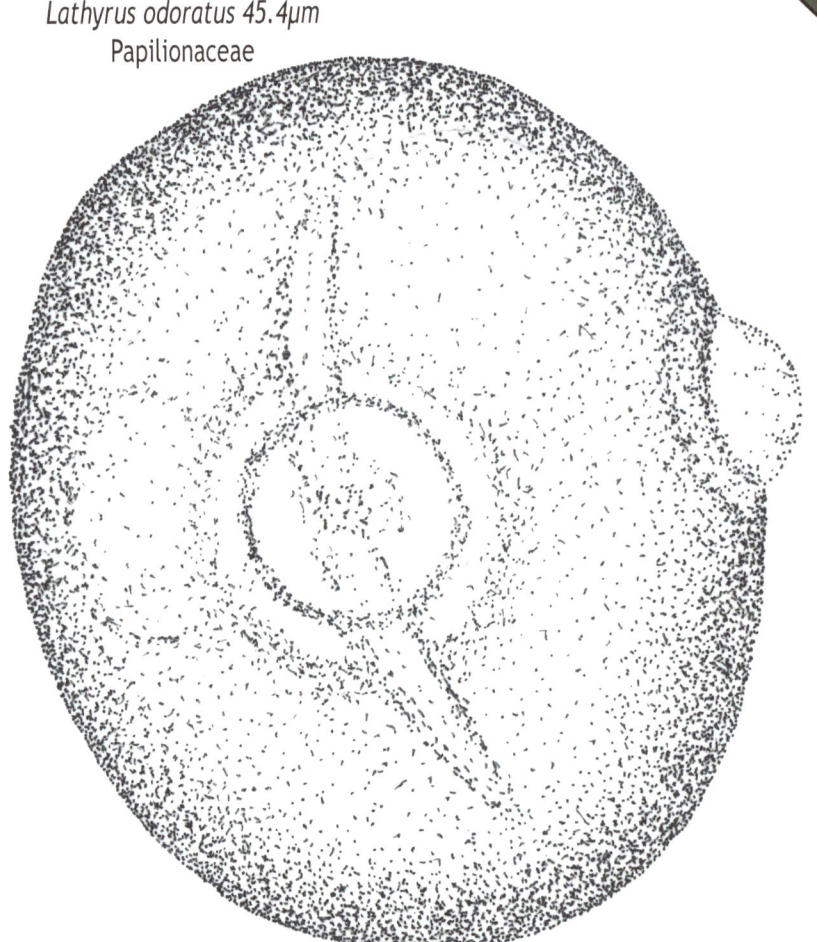

Highly scented annual climber, popular as a cut flower. The flowers self pollinate while the buds are still closed.
The pollen grain is slightly elongated in equatorial view, with three furrows, each having a pore at the centre.

Sycamore

Acer pseudoplatanus 33.8μm
Sapindaceae

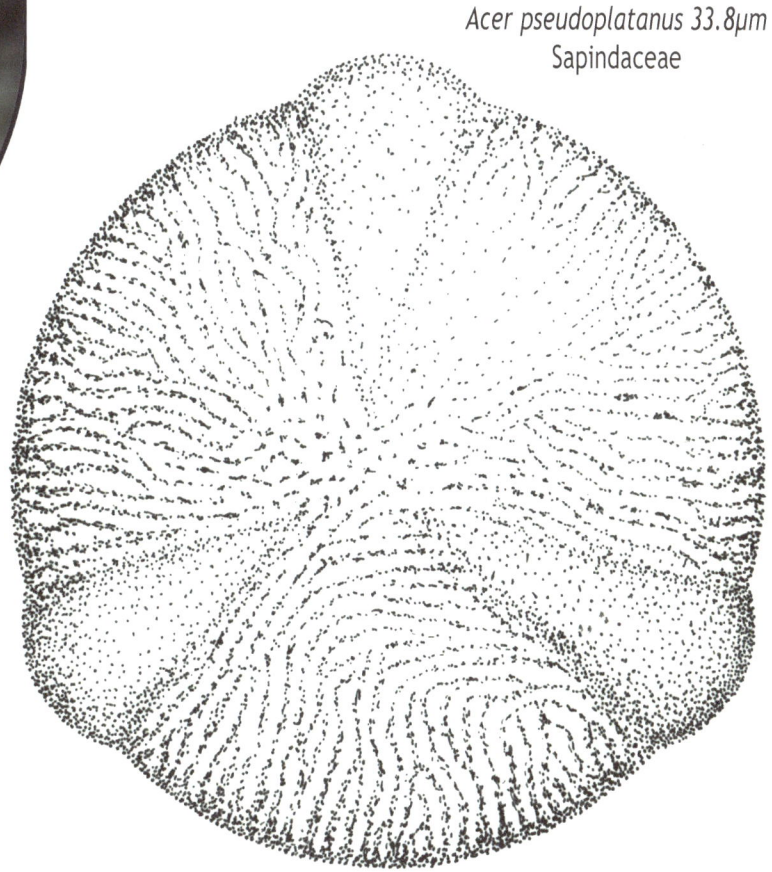

The trees can grow tall and live long. The monoecious racemes appear in spring. They are visited by insect pollination vectors with pollen dispersal also by the wind. It is very effective at seed propagation with seedlings sprouting copiously in the vicinity and growing rapidly.
The surface of the pollen grain has a longitudinally striated pattern.

Teasel

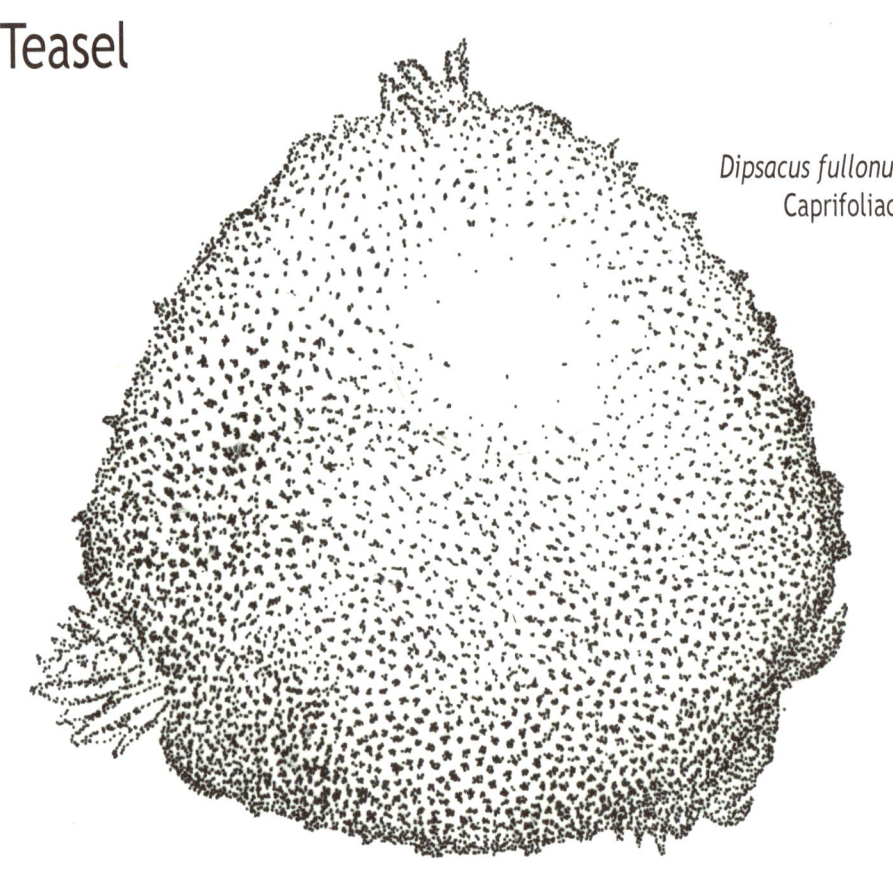

Dipsacus fullonum 32.0μm
Caprifoliaceae

Tall, distinctive, prickly biennial wild flower, a rich food source for insects and when the seeds are produced, for small birds.
The pollen grain has a dotty surface and tufty protrusions from its three pores.

225
Meadow thistle

An erect perennial thistle preferring a damp boggy habitat.

Cirsium dissectum 56.3μm
Asteraceae (Compositae)

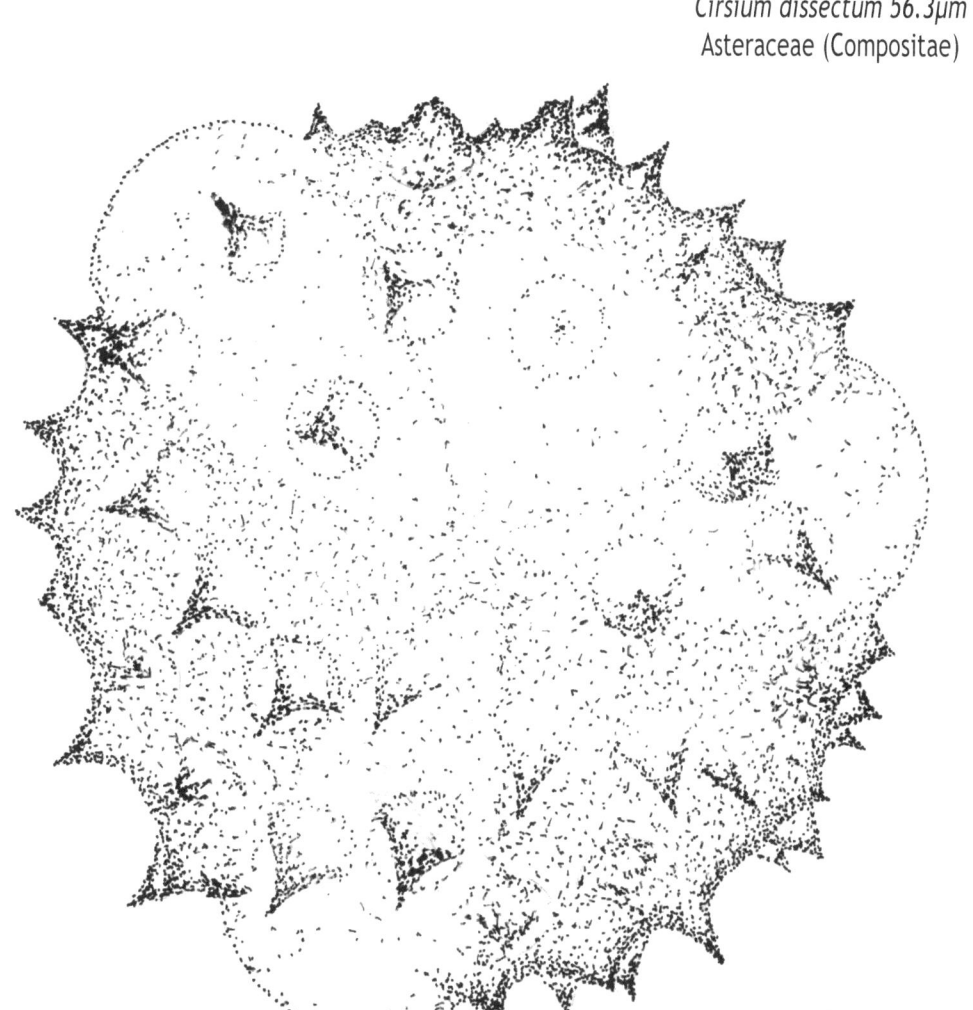

Marsh thistle

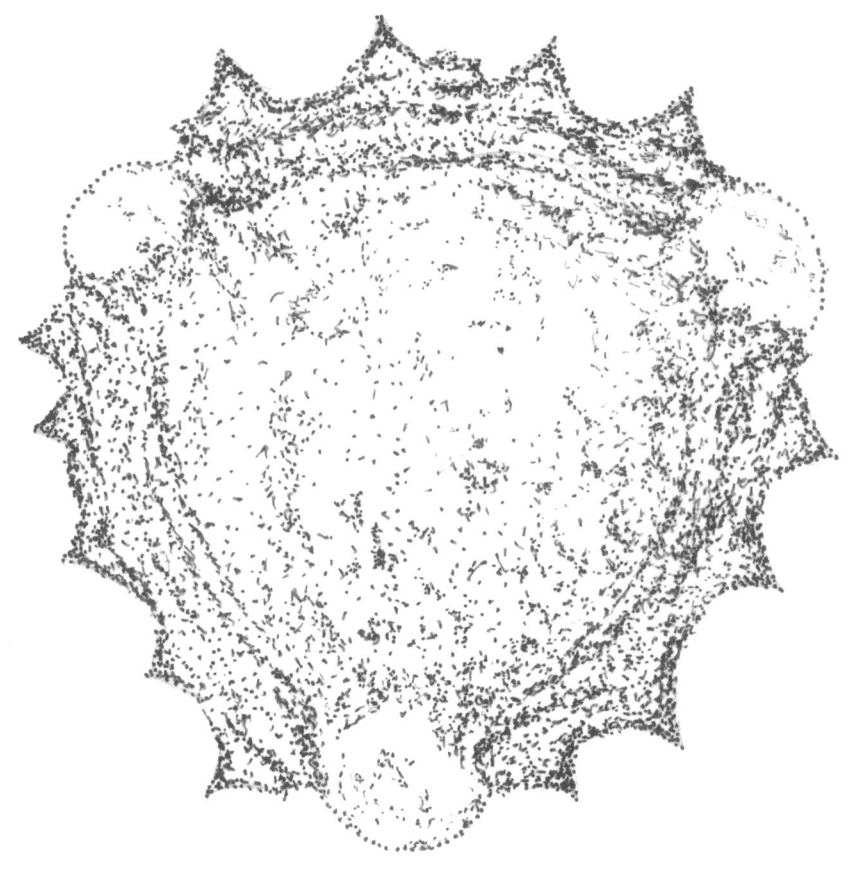

Cirsium palustre 37.4μm
Asteraceae (Compositae)

Tall thistle common in damp areas. It
produces a lot of nectar.
The pollen grain has three short furrows
almost entirely taken up by a large pore
at the centre.

Spear thistle

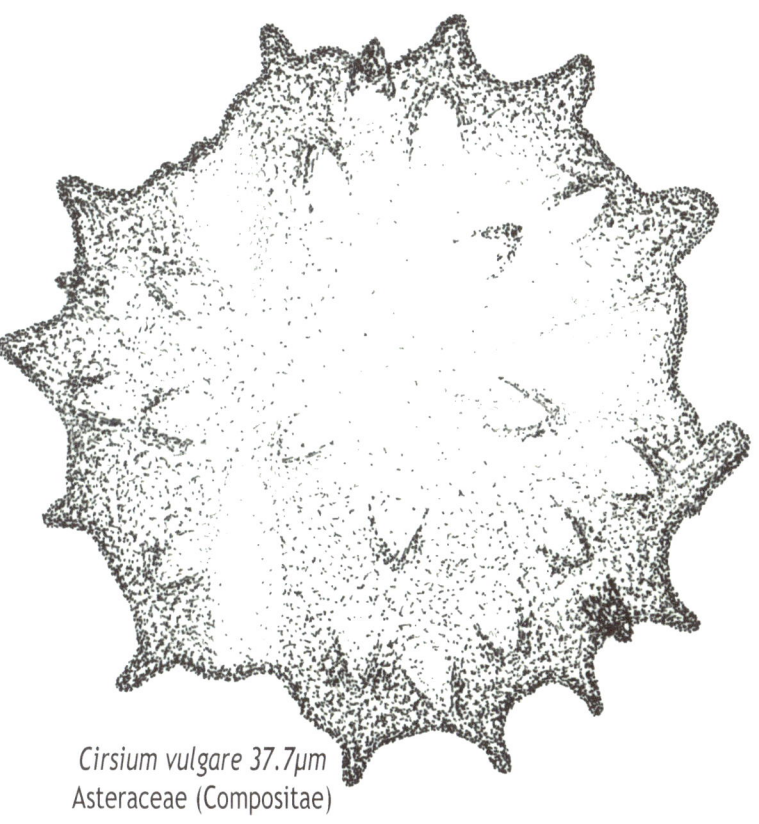

Cirsium vulgare 37.7μm
Asteraceae (Compositae)

A tall, short-lived perennial, the common thistle also produces lots of nectar for pollinators.
The pollen grain has three furrows and the spiny surface characteristic of the Asteraceae.

Yellow melancholy thistle

Cirsium erisithales 30.6μm
Asteraceae (Compositae)

Polar view of the grain showing the typically spiny surface of the pollen grain which has three short furrows each with a large pore.

Creeping red thyme

Thyme is attractive to bees and other insects. The pollen grain has seven furrows.

Thymus sp. 31.1µm
Lamiaceae

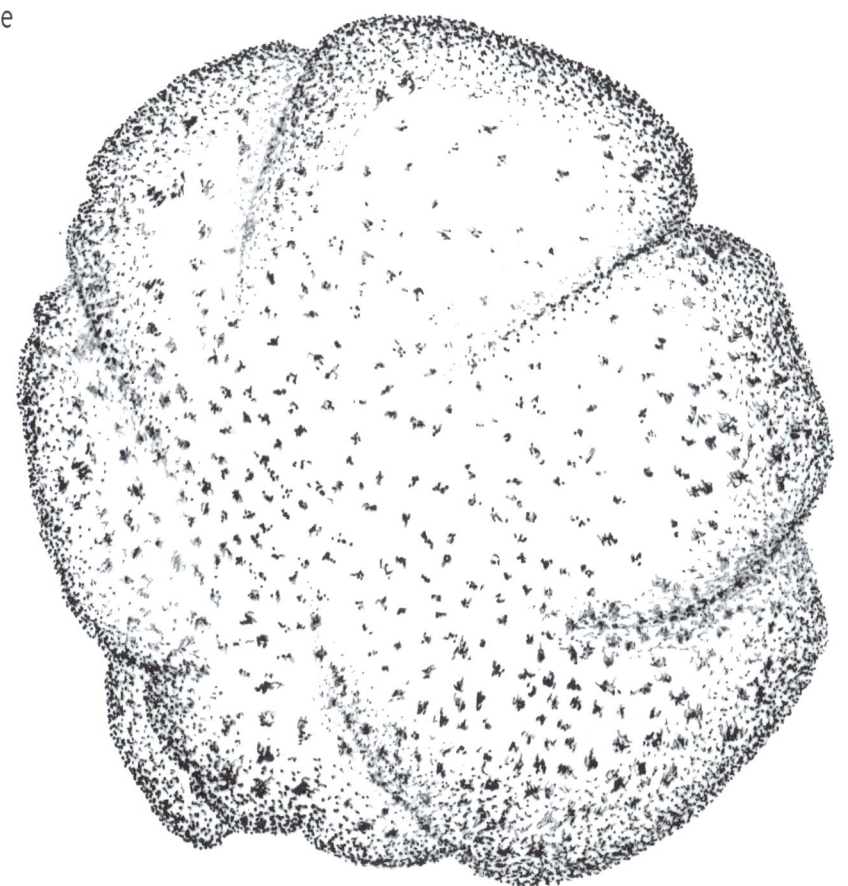

Annual thymelaea (Passerina)

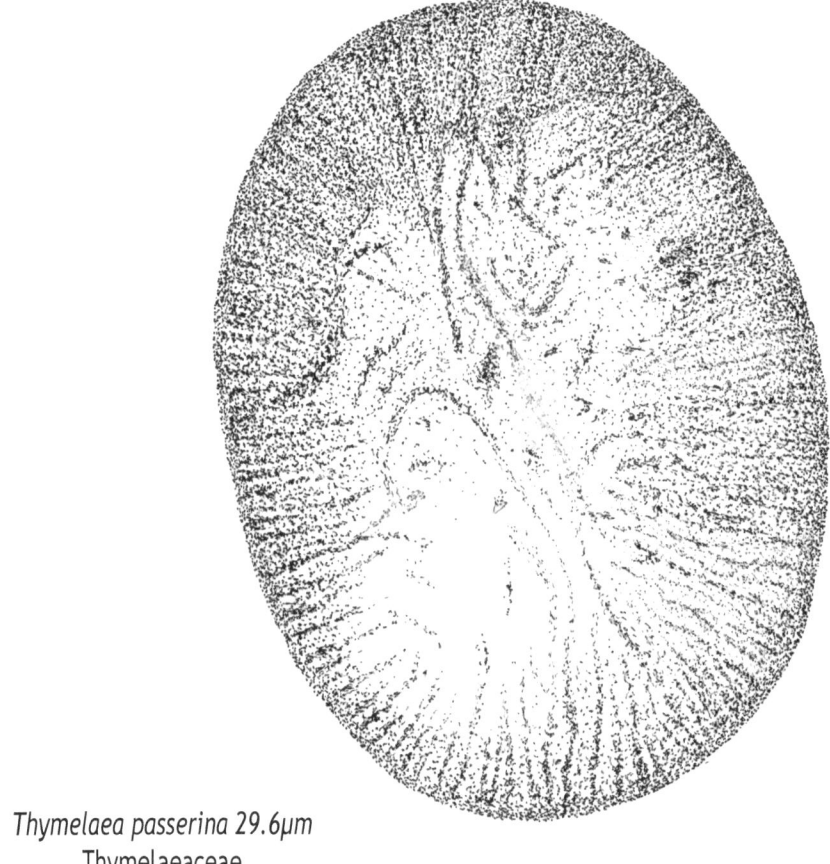

Thymelaea passerina 29.6µm
Thymelaeaceae

A rather insignificant looking plant with a growth habit similar
to heather and having small flowers. It is wind pollinated.
The pollen grain is spheroidal. It has a reticulated surface
scattered with several not very obvious pores at regular
intervals over the surface.

Common toadflax

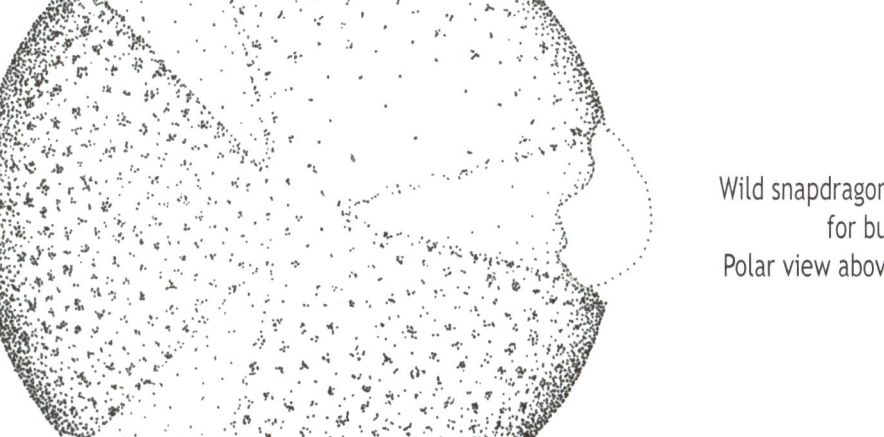

Wild snapdragon, with flowers similarly adapted for bumble bee pollination.
Polar view above left, transparent visualisation below left.

Linaria vulgaris 18.1μm
Plantaginaceae

Tomato

The tube at the centre of the tomato flower is made from five stamens, which yield their pollen when vibrated. Bumble bees are able to access the nectar, and provide the vibration, enabling them to carry pollen from flower to flower. Their activity results in greater crop yields. The pollen grain has three long furrows each with a pore at the equator. It has a dotted pattern over its surface, and to a lesser extent along the furrows as well.

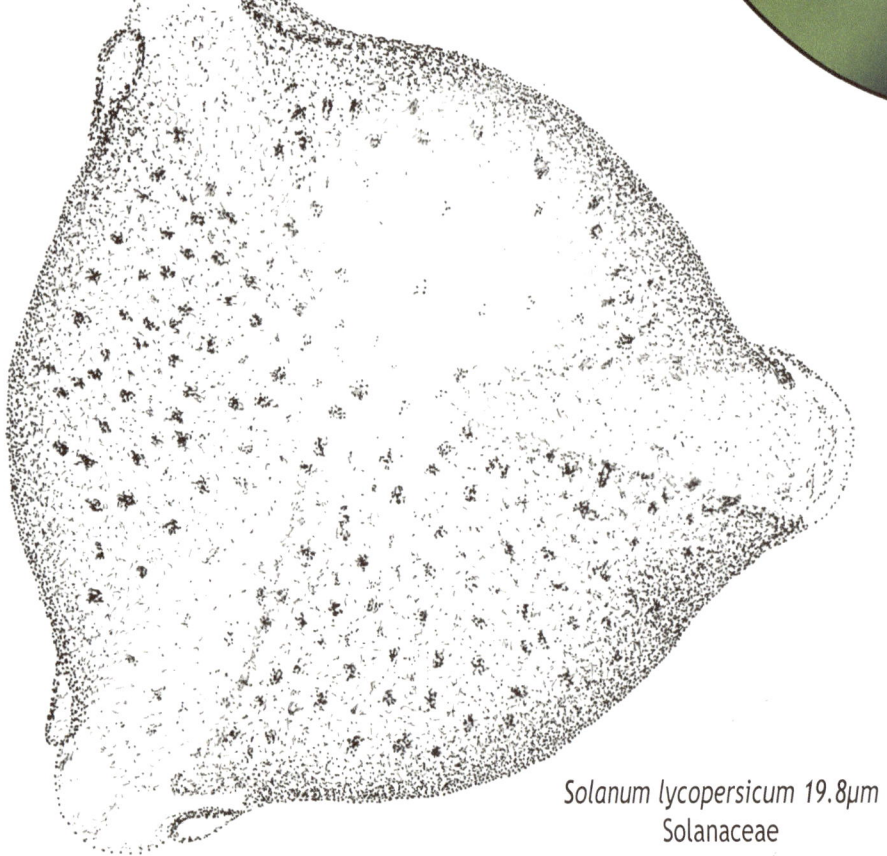

Solanum lycopersicum 19.8μm
Solanaceae

Tulip

Pollen grains from the tulip family are eliptical with one aperture.

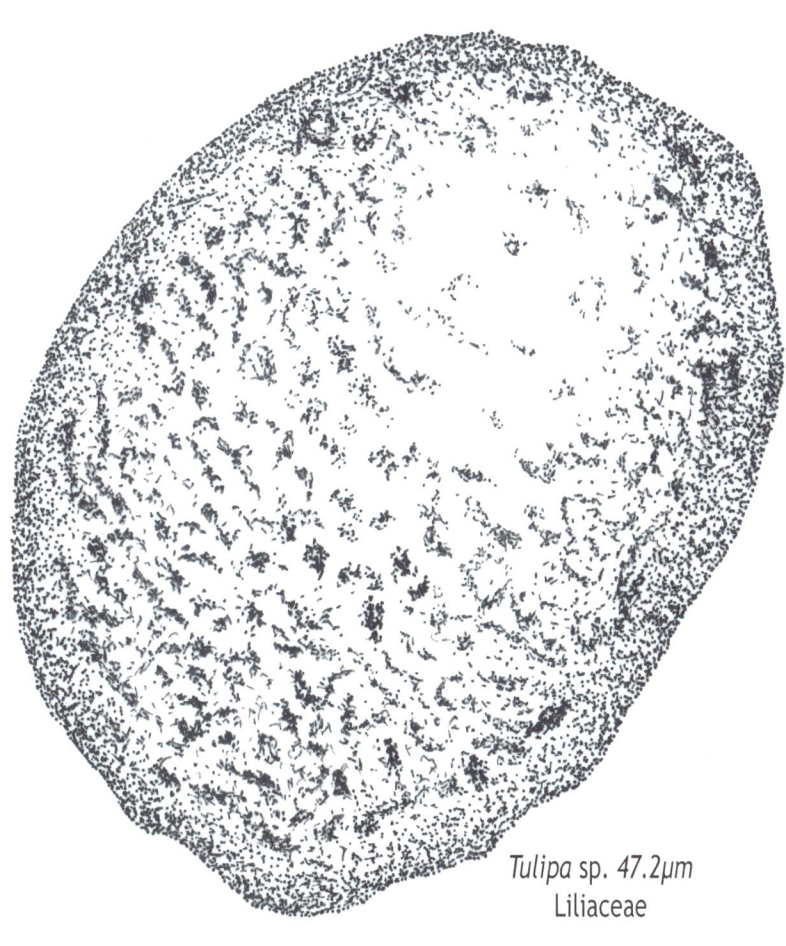

Tulipa sp. 47.2μm
Liliaceae

Red valerian

Centranthus ruber 64.0μm
Caprifoliaceae

Photo © Sally Dunn

A common perennial plant with showy clusters of profuse, small individual flowers. It is a good source of nectar for bees, butterflies and moths. The pollen grain exine shows a knobbly surface on the furrows.

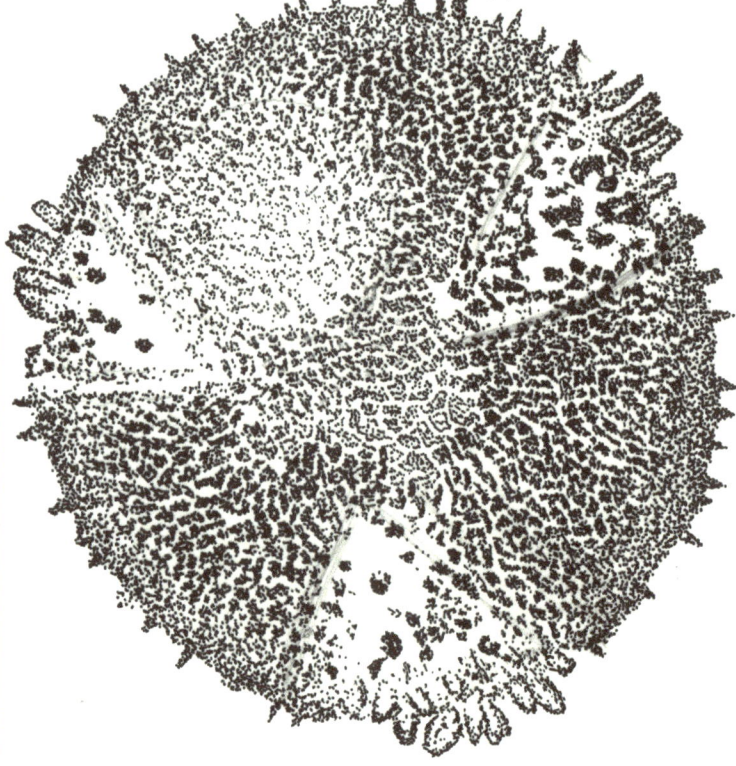

Verbena x hybrida

Drought resistant species and hybrids attractive to butterflies. The pollen grain is triangular with wide, decorated furrows with protruding pores.

Verbena x hybrida 50.6µm
Verbenaceae

Tufted vetch

A scrambling plant seen in woodland, scrubland and grassland, tufted vetch has a long flowering period late spring to late summer, and is a good source of nectar for bees and butterflies. Its elongated pollen grain is shown in equatorial view, having three short furrows each with a prominent pore at the centre.

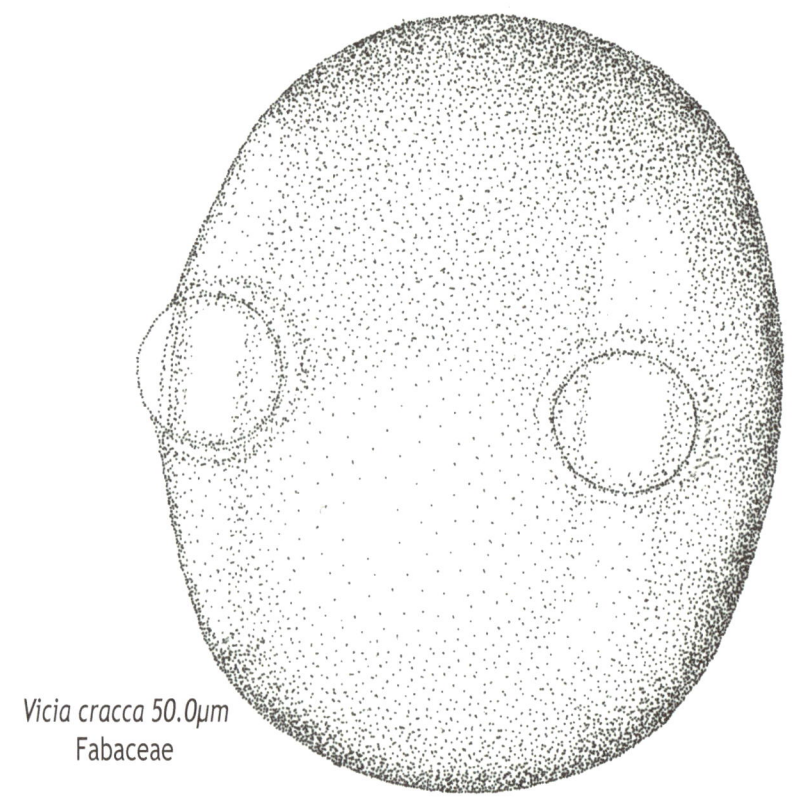

Vicia cracca 50.0μm
Fabaceae

Viburnum

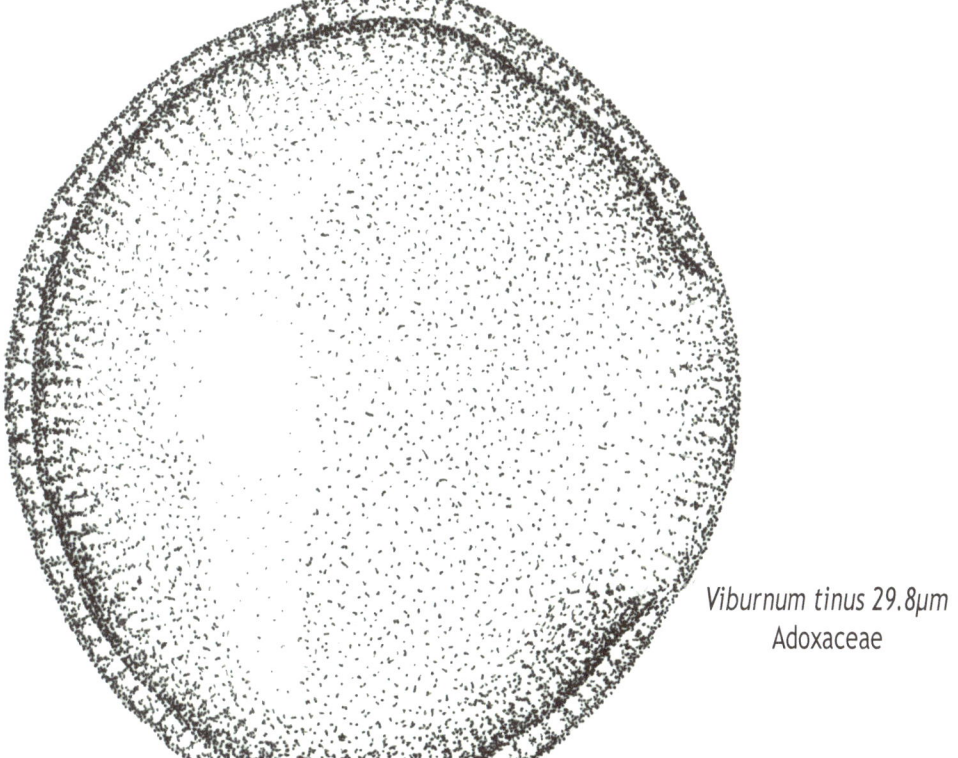

Viburnum tinus 29.8µm
Adoxaceae

An evergreen shrub flowering in late winter, early spring. The pollen grain in equatorial view has three furrows each with a bulging pore. The surface has a lattice netting.

Sweet violet

A low growing, perennial, spreading plant, with sweetly scented flowers. This polar view of the pollen grain shows three long wide ornamented furrows. Each has a pore at the equatorial centre.

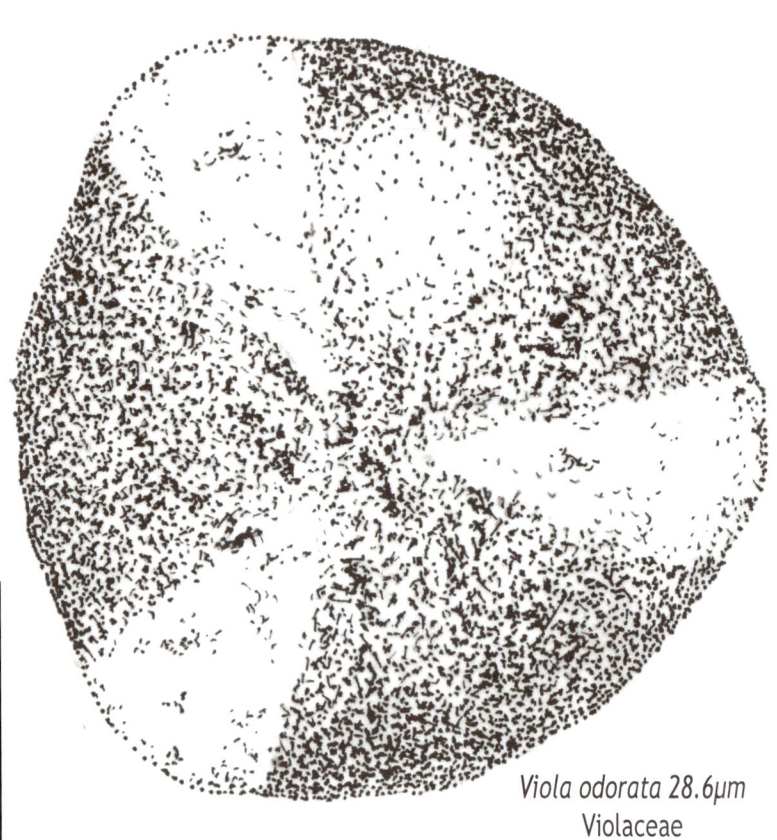

Viola odorata 28.6µm
Violaceae

Perennial wall-rocket

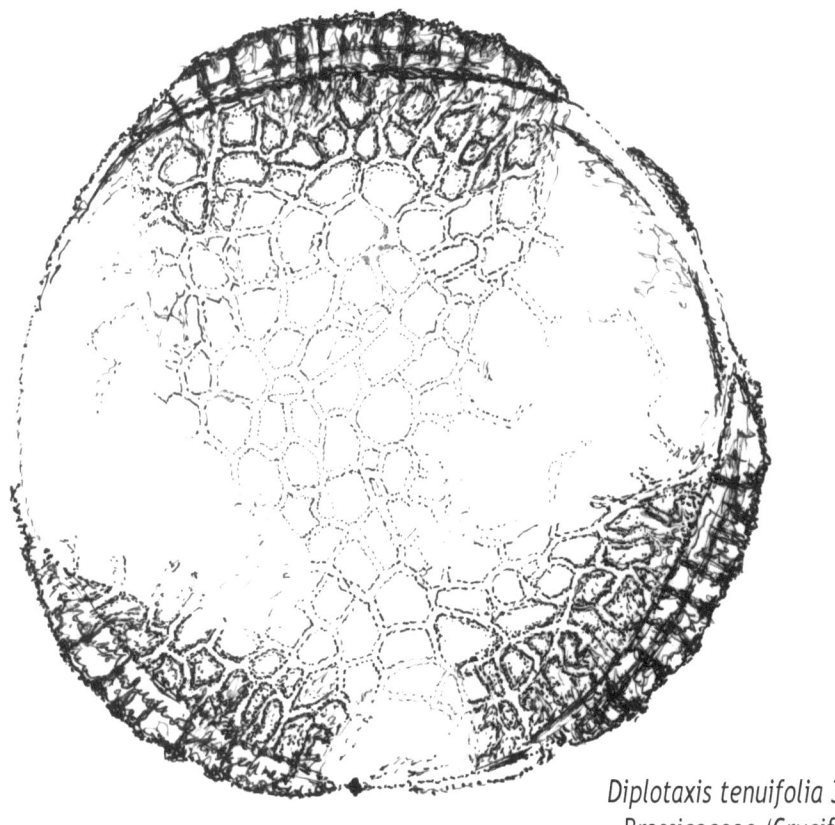

Diplotaxis tenuifolia 34.0µm
Brassicaceae (Cruciferae)

A member of the mustard family, its leaves have a strong flavour and together with cultivated varieties, are popular in salads.
The pollen grain has three slightly indented, decorated furrows and a netting pattern over the rest of the surface.

Walnut

A large tree renowned for both the edible nuts and attractively grained wood. It is monoecious with separate male and female catkins on the same tree. The male flowers are drooping yellow-green catkins 5-10 cm long. The female flowers appear in clusters of 2-5, are pollinated by wind, and develop into the green capsuled nut.

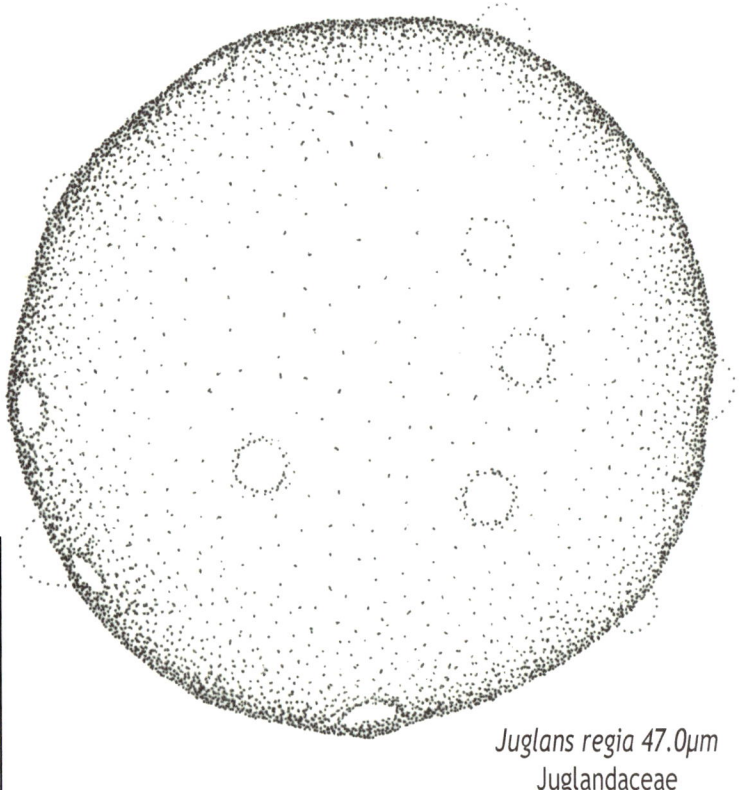

Juglans regia 47.0μm
Juglandaceae

The pollen grain is quite large for a wind pollinated grain, spherical and covered in pores.

Photos © Sally Dunn

Wayfaring tree

A small tree or shrub with fragrant flowers produced in May to June. The resulting autumn berries are attractive to birds and small mammals but poisonous to humans. The pollen grain has three furrows and a bobbly reticulate surface.

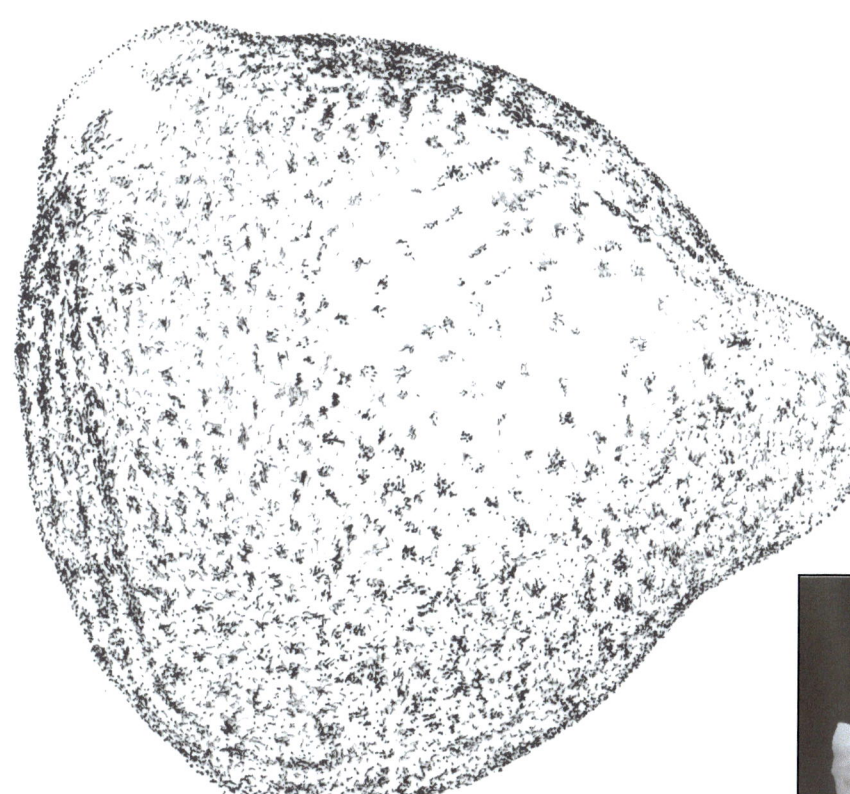

Viburnum lantana 28.3μm
Adoxaceae

Weld

In former years, what is now a weed was used to produce a yellow dye. The pollen grain is smaller than average with three furrows, each mostly taken up with a large pore.

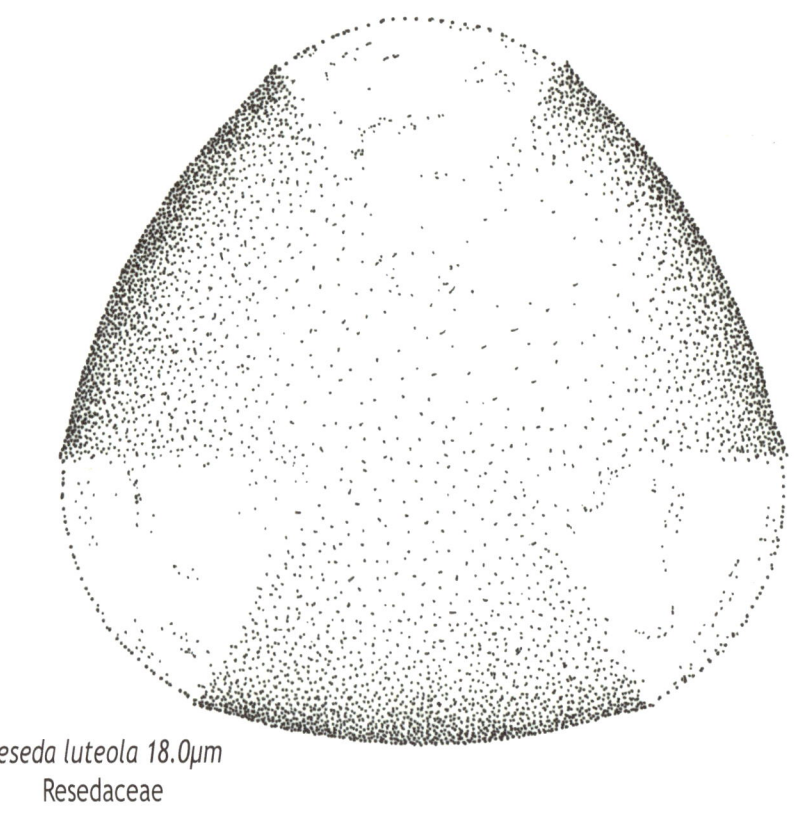

Reseda luteola 18.0µm
Resedaceae

Crack willow

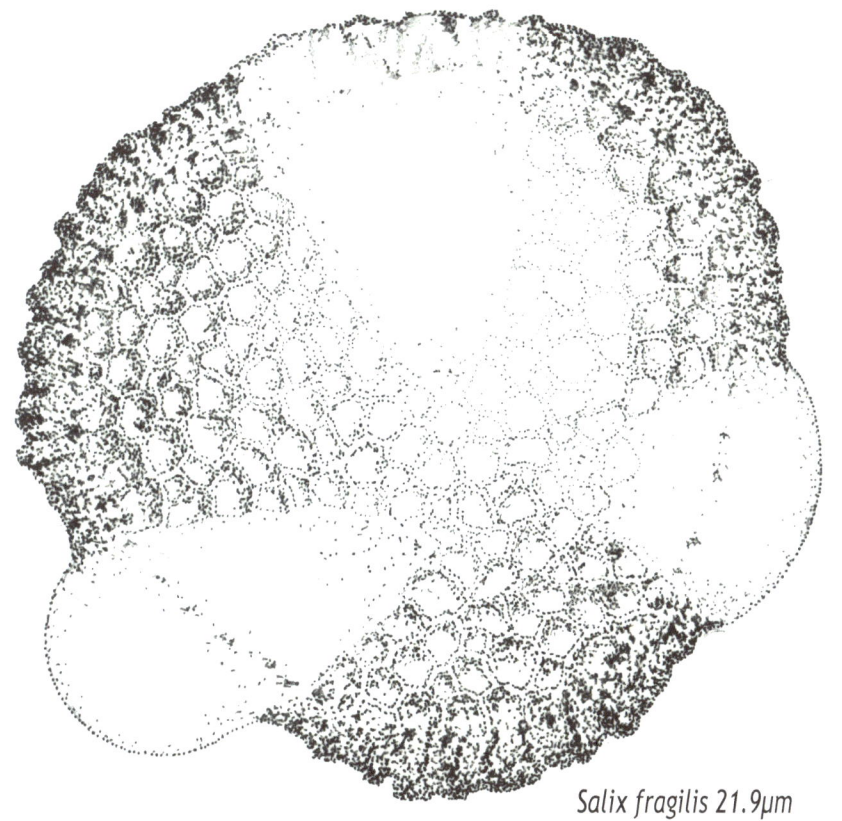

Salix fragilis 21.9µm
Salicaceae

Fast growing tree, dioecious with long catkins in May, female being green and male catkins yellow. The catkins have nectaries, are insect pollinated and the seeds later dispersed by wind. This willow gets its name from twigs breaking easily with an audible crack. Broken twigs take root easily where they come to rest. As willows often grow along riverbanks twigs can be carried for some distance along the river. The pollen grain, illustrated from the pole, is spheroidal, has a reticulated surface with a regular pattern, and three wide furrows.

Crack willow and white willow catkins look very similar. Also the trees readily hybridise: cricket bat willow is thought to be a hybrid of these two.

Goat willow

The familiar pussy willow, dioecious with yellow male catkins and green female catkins on different trees. They provide an early source of pollen and nectar for bees and other insects.
The pollen grain is spheroidal with three wide, deep furrows and a reticulate pattern over the surface between furrows.

Salix caprea 17.6µm
Salicaceae

Grey willow

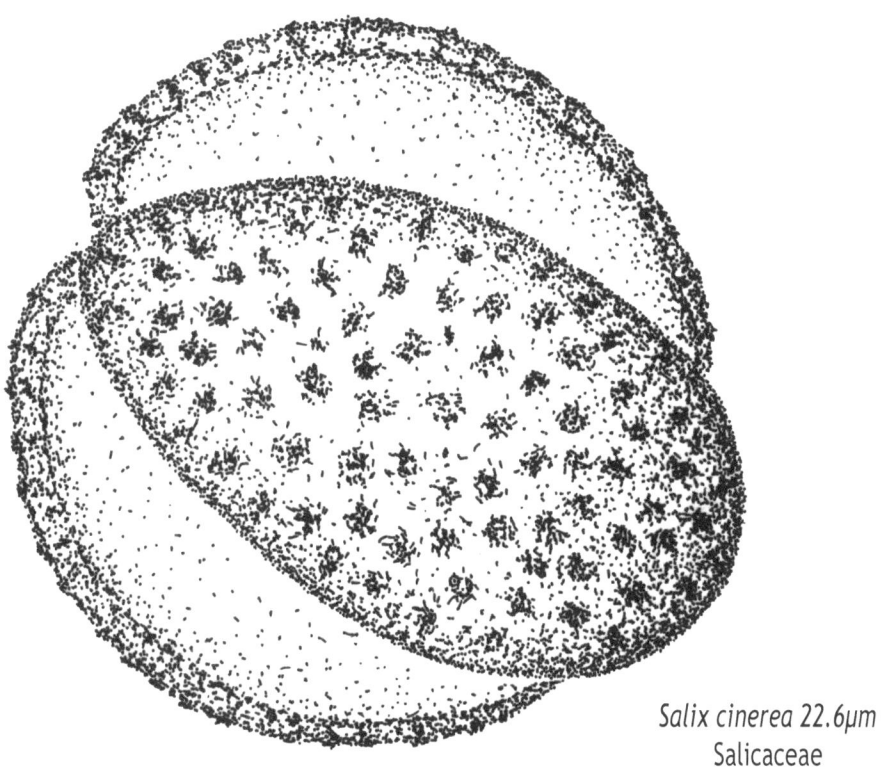

Salix cinerea 22.6μm
Salicaceae

Unlike the thin leaves of most willows, the leaves of grey willow are oval, twice as long as wide. Closely related to Goat willow, the two are not easy to distinguish, and also hybridise readily. Their furry catkin buds are similar. Their pollen grains are also similar in appearance, spheroidal with three wide, furrows and a reticulate pattern over the surface between furrows.

Broad leaved willowherb

The broad leaved willowherb is a food plant for the larvae of several species of moth.
Single flowers are seen at the end of long, drooping flower tubes.
Polen grains aggregate in a tetrad tetrahedral formation with with pores aligned.

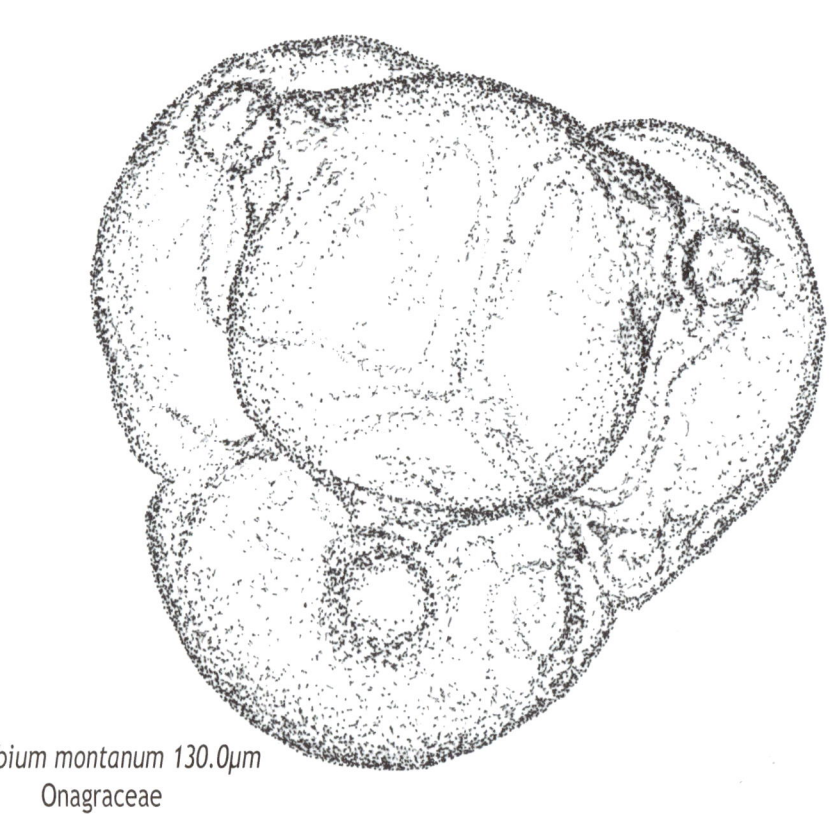

Epilobium montanum 130.0µm
Onagraceae

Great willowherb

Great willowherb is a tall, hairy perennial found in damp habitats. It has deep magenta-pink flowers July to August followed by fluffy seeds. The pollen grains form large polyad clusters of four grains. Note the microscopic hairs.

Epilobium hirsutum 137.0µm
Onagraceae

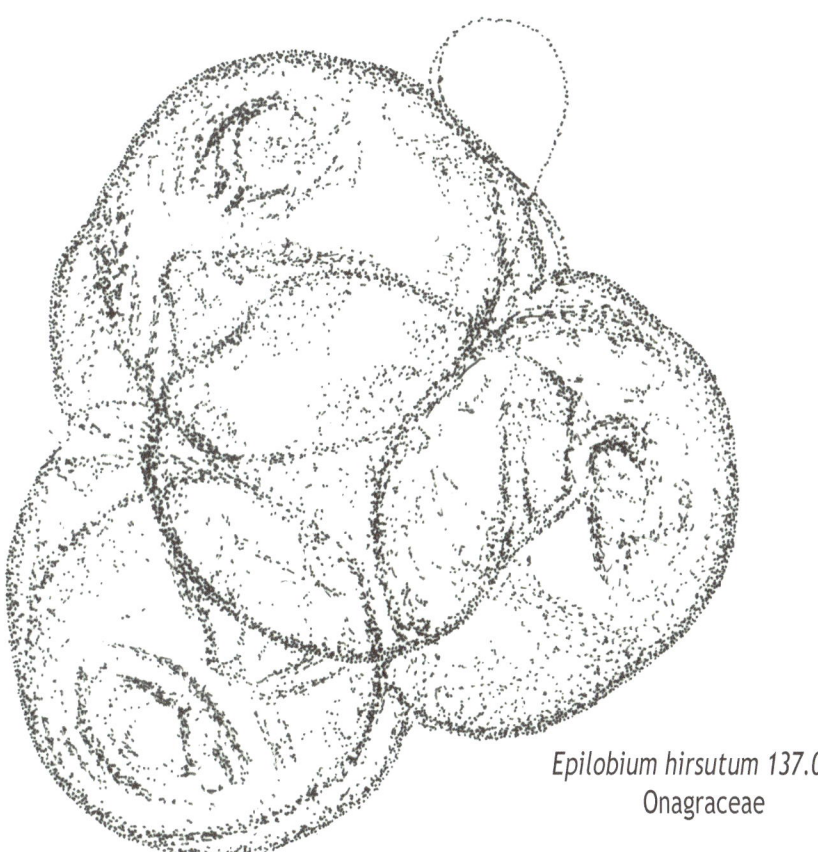

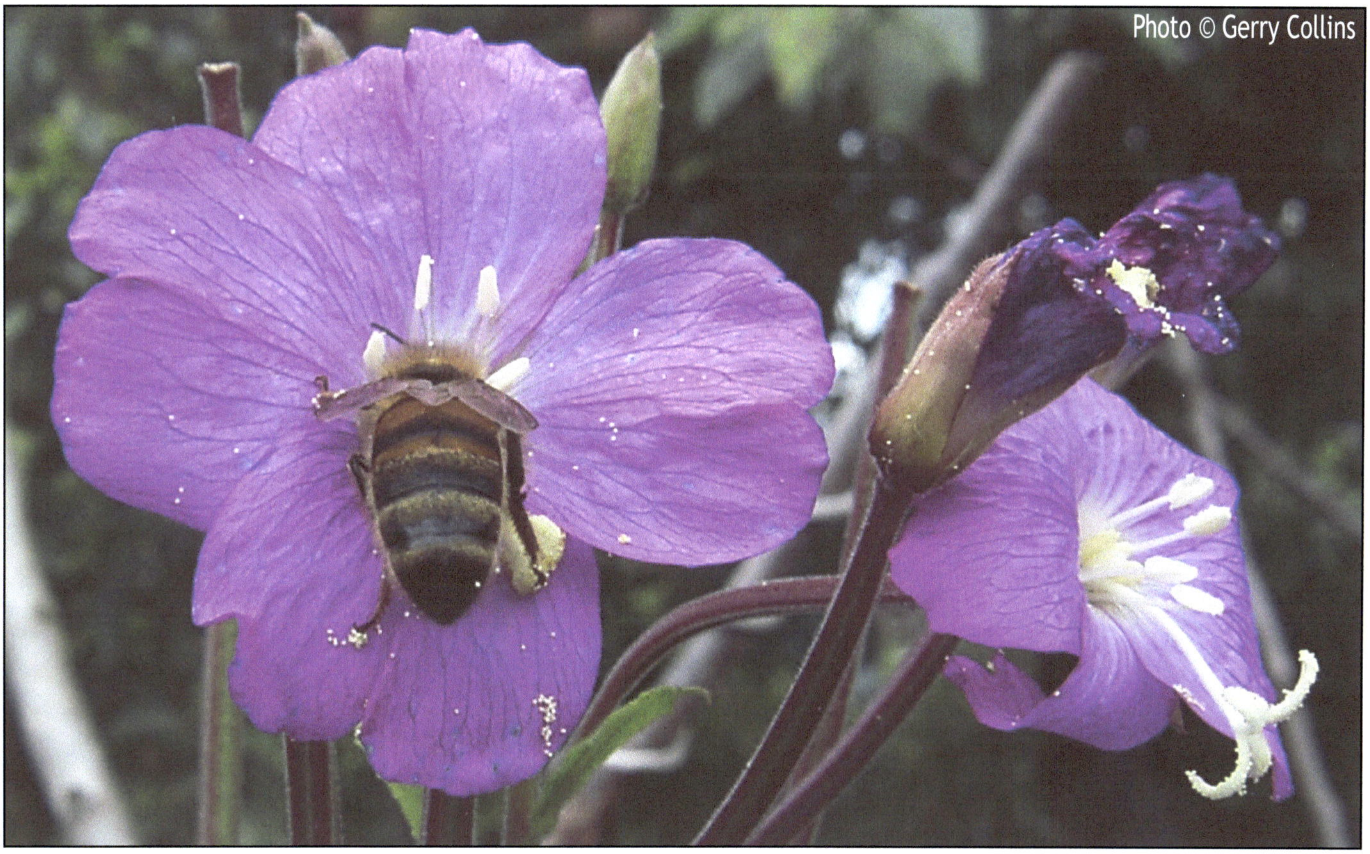
Photo © Gerry Collins

Rosebay willowherb

Photo © Janet Morris

Seen as distinctive and striking swathes of reddish purple, it is also known as fireweed, being one of the first to seed itself in grounds following clearance by fire. The anthers develop ahead of the ovary, a common method to avoid self pollination and encourage cross pollination. It is a food plant for the larvae of the elephant hawk moth.
The larger than average pollen grains don't cluster in pyramids. They have three pores and characteristic straggling, hairy extensions. The pollen is blue.

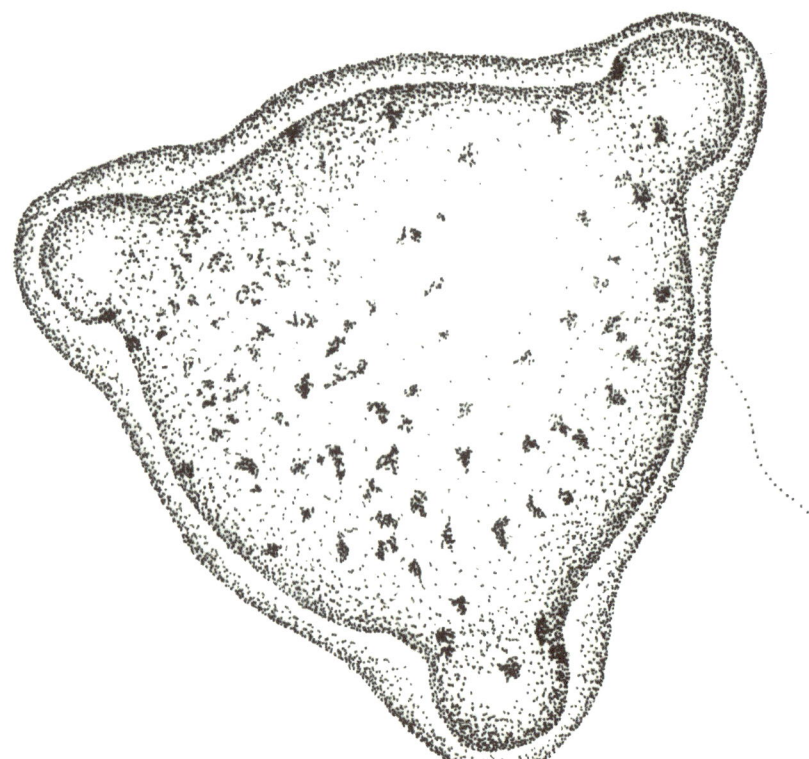

Chamaenerion angustifolium (Epilobium angustifolium) 76.0μm
Onagraceae

Common wintercress

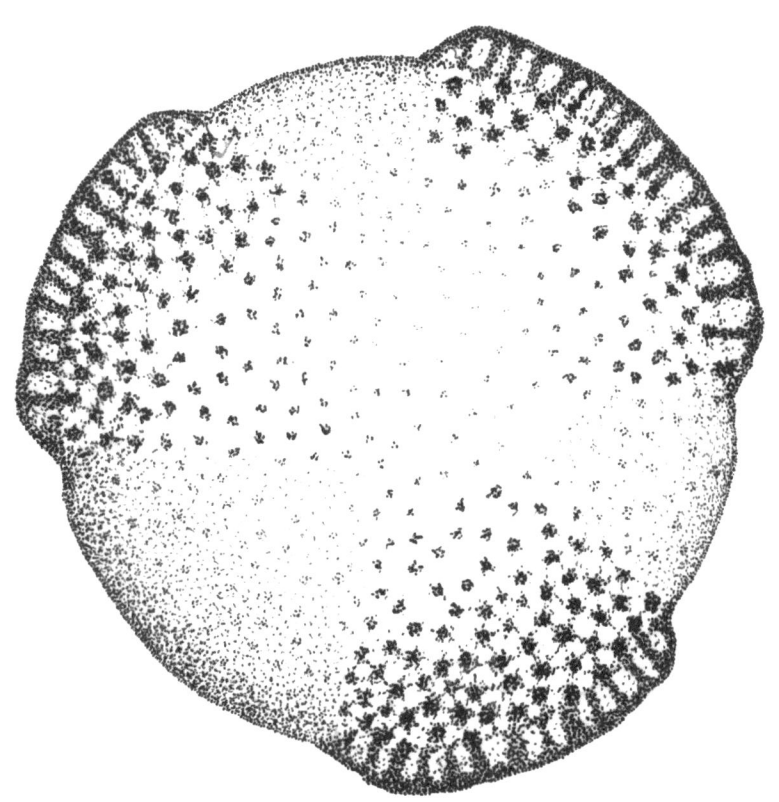

Barbarea vulgaris 22.0μm
Brassicaceae (Cruciferae)

A common, invasive, recognisably cabbage family, edible winter weed with small yellow flowers.
The pollen grain is spheroidal with a reticulate surface between the three quite wide furrows.

Winter flowering jasmine

As the name suggests, this untidy bush flowers between November and March, providing nectar and pollen for insects able to forage at a time of year when food is scarce.
The pollen grain has a reticulated surface and three furrows.

Jasminum nudiflorum 44.0µm
Oleaceae

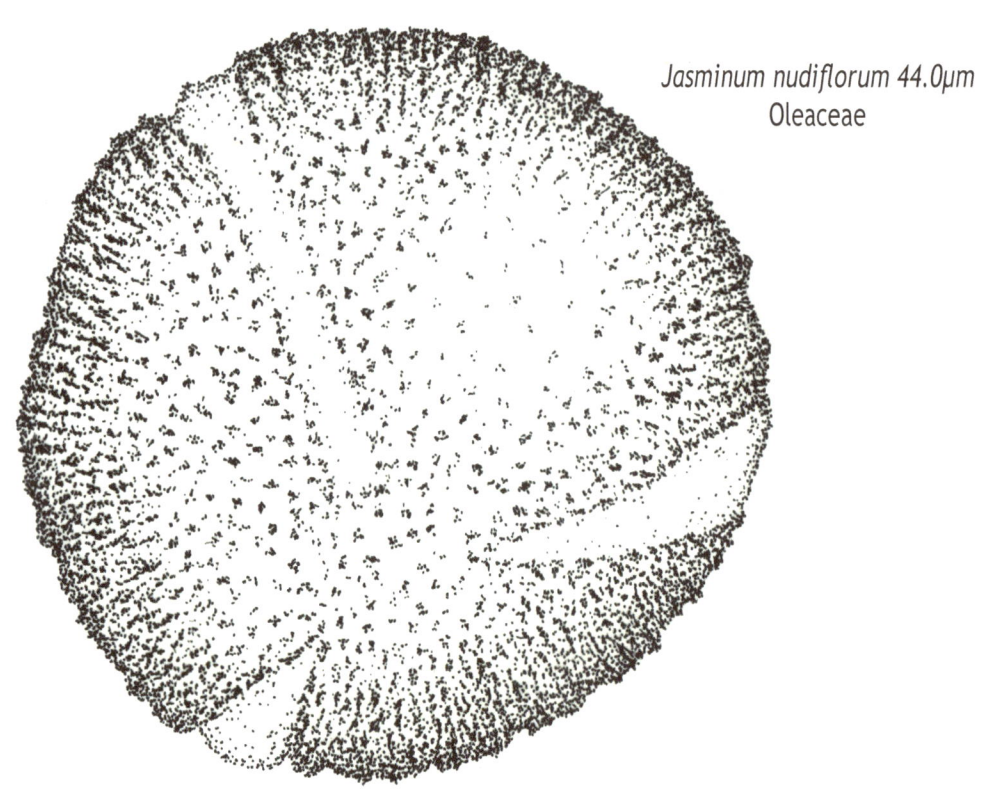

Wintersweet

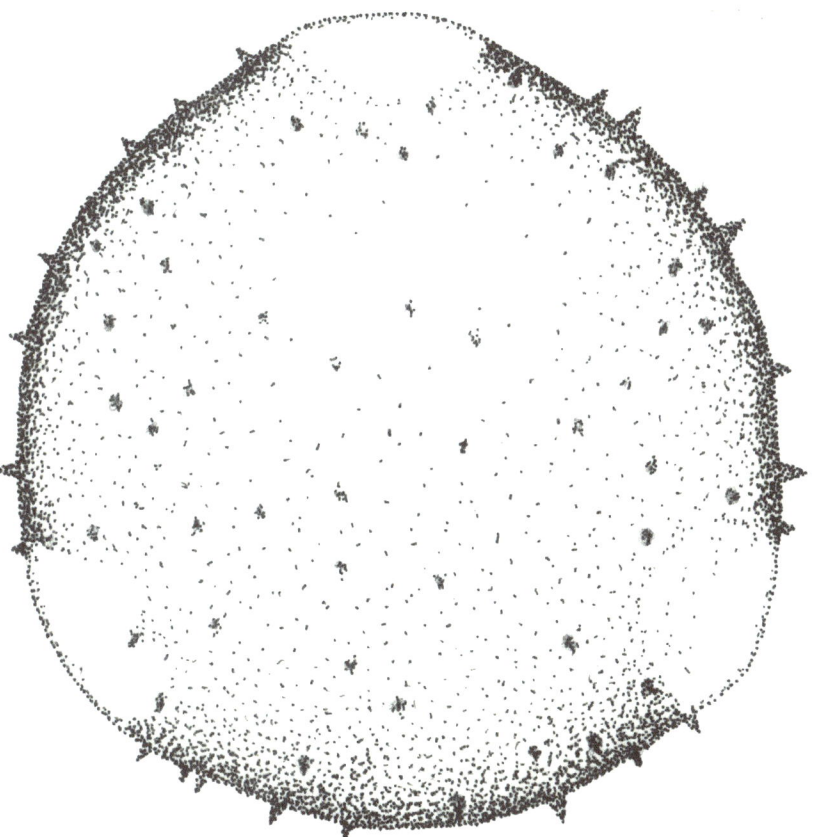

Chimonanthus praecox 60.6µm
Calycanthaceae

The strongly scented flowers are produced in late winter and are pollinated by insects.
The pollen grains are larger than average.

Wood sorrel

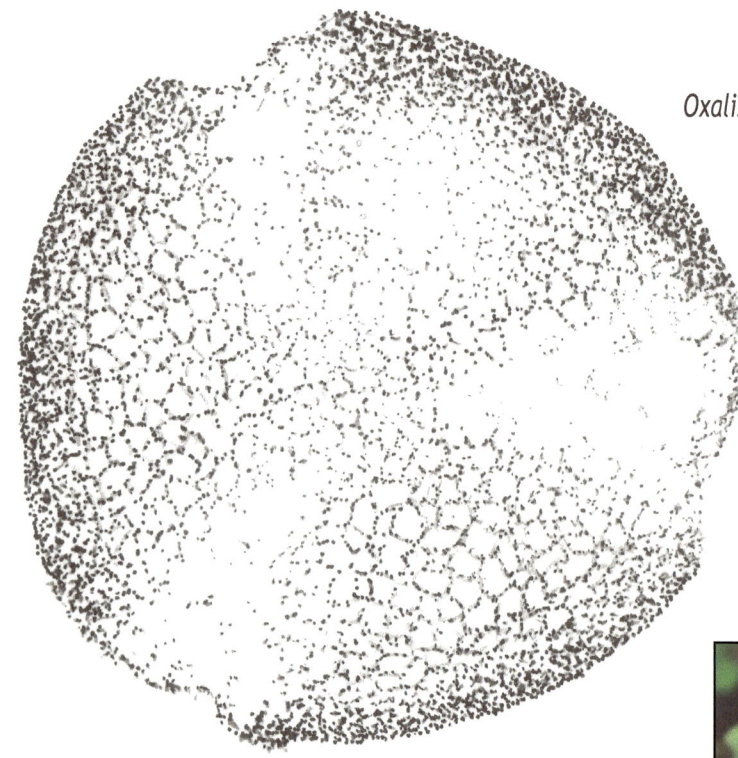

Oxalis acetosella 26.0μm
Oxalidaceae

A low growing, spreading, woodland plant with shamrock like leaves and white flowers.
The pollen grain surface has a close reticulated pattern and furrows decorated with protruding dots.

Yarrow

The flat topped cluster of this wild flower is visited by many insects.

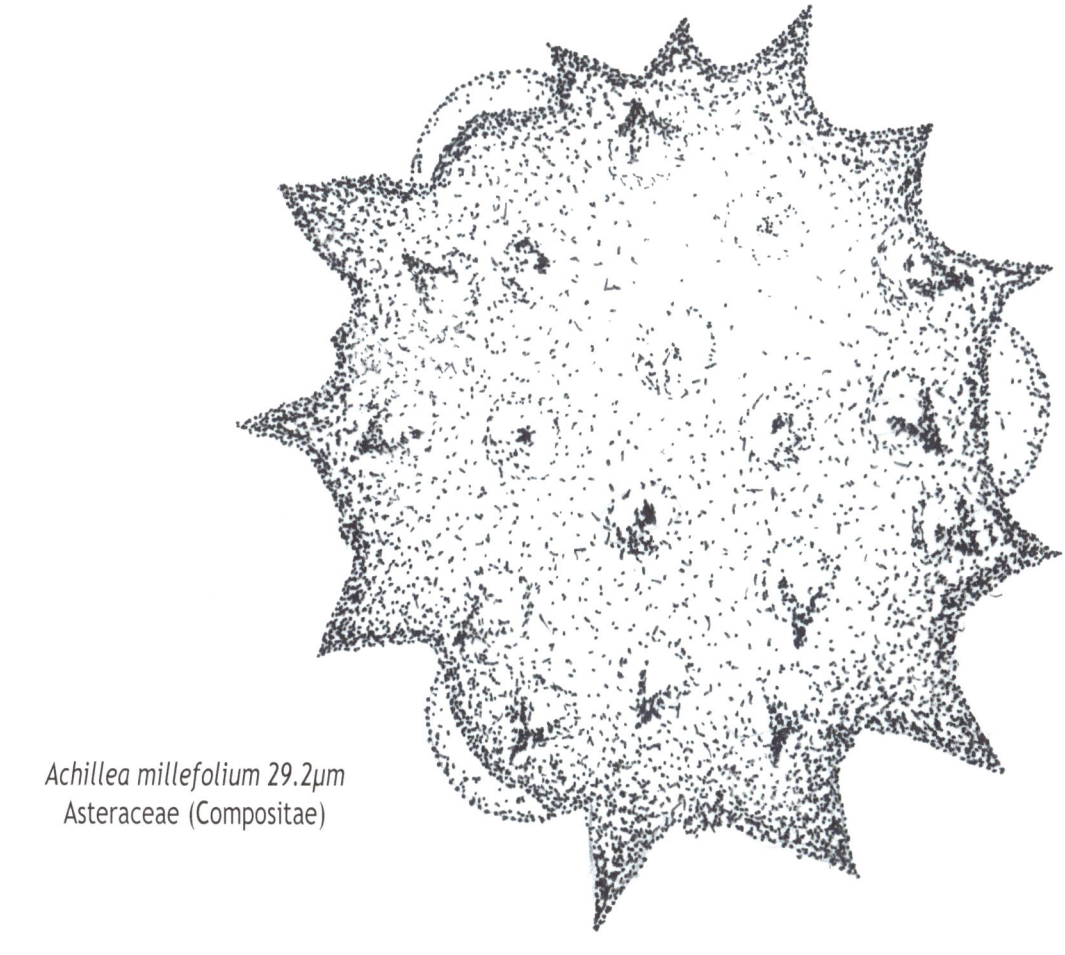

Achillea millefolium 29.2μm
Asteraceae (Compositae)

Yellow archangel

This lamium pollen has three furrows with a reticulate surface pattern visible under the light microscope.

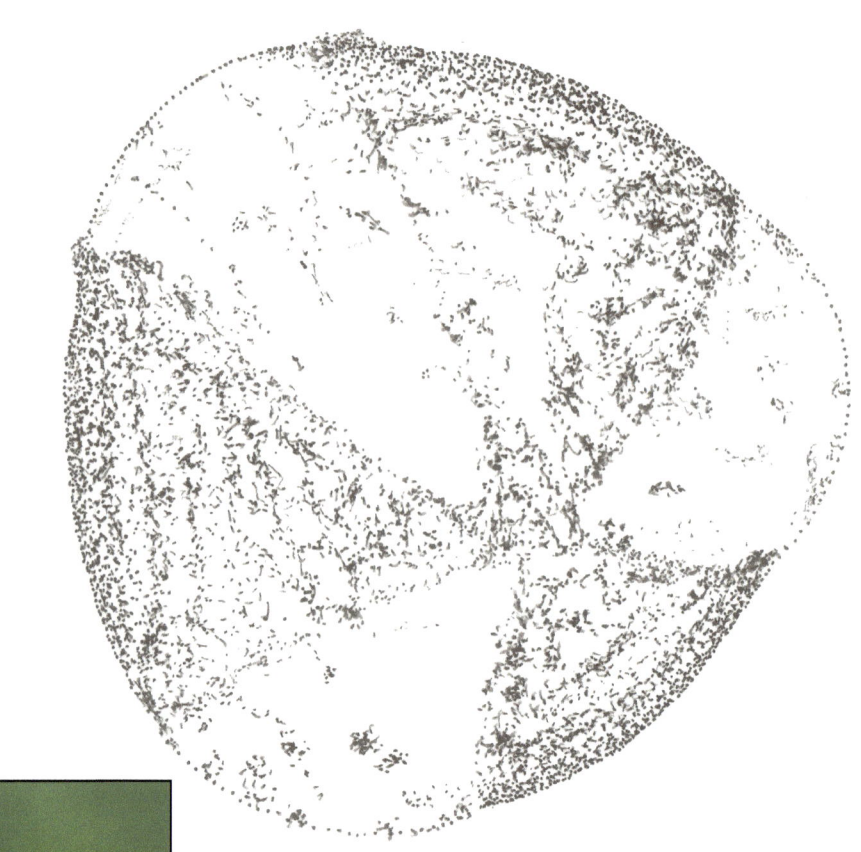

Lamium galeobdolon 27.0μm
Lamiaceae

Photo © Sally Dunn

Yellow rattle

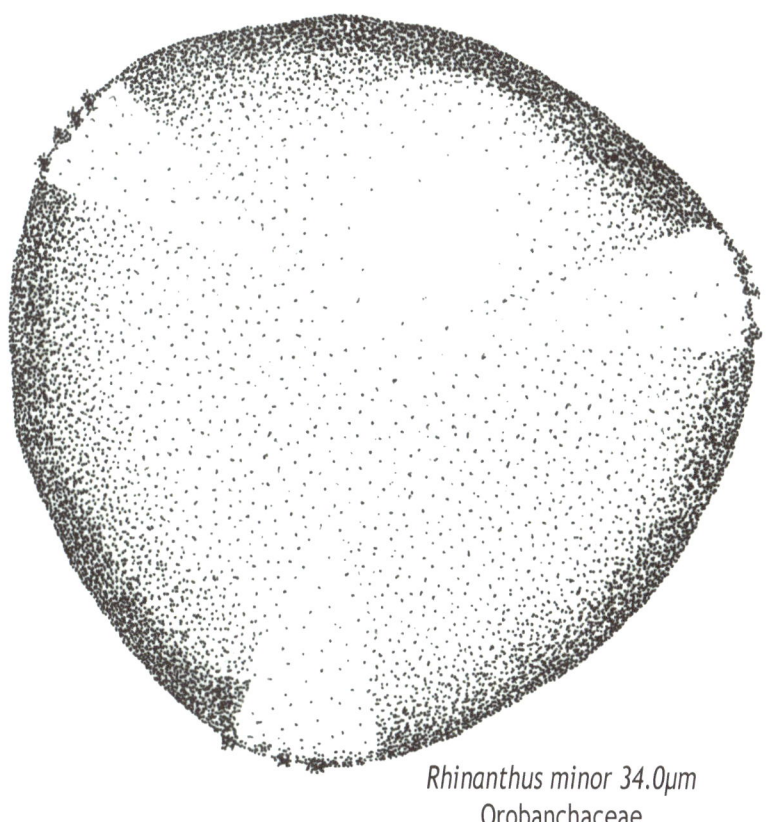

Rhinanthus minor 34.0µm
Orobanchaceae

A meadow flower, the seeds rattling when
shaken in the pods giving it its name.
The pollen grain's three decorated furrows
converge towards the poles.

Yew

Can be used for hedging, but left undisturbed can grow to a huge evergreen tree. The female trees bear berries much loved by thrushes but poisonous to animals including humans. The male trees produce copious dustings of pollen, dispersed by the wind in visible drifts around the end of February. The pollen grain is hard to distinguish from buckthorn pollen.

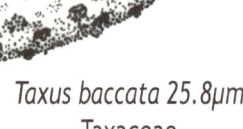

Taxus baccata 25.8µm
Taxaceae

Acknowledgements

Many more people, plants, insects, and in some cases other creatures have contributed to the second edition of this book.
Many thanks to all of you, and particularly the photographers,
and all who have given their time to advising, checking and proof reading.
For the latter, thanks especially to Dr Sally Dunn, Andrew Halstead, Janet Morris, and Dr Eva Wallender, Sweden.
It has been great fun and some times a greater quest tracking down some interesting pollen and adding the
corresponding photographs at the plants' flowering times.
We'd like to thank Tregothnan Estate in Cornwall, The Royal Horticultural Society and their Gardens at Rosemoor, Devon
and several other gardens, parks and flower shows, in fact all willing and unwitting contributors.

References

1. Pierre Binggeli & James Power: Gender variation in ash (*Fraxinus excelsior* L.)
   http://www.mikepalmer.co.uk/woodyplantecology/docs/MNR-ashgender.pdf

2. Hodges, D. (1952) The pollen loads of the honey bee.
   Reprinted 1984: International Bee Research Association

3. Kirk, W.D.J (1994) A Colour Guide to Pollen Loads of the Honey Bee.
   Revised edition 2006: International Bee Research Association

4. Plants having extra floral nectaries http://www.extrafloralnectaries.org/the-list.html

5. Pollen images http://www-saps.plantsci.cam.ac.uk/pollen/

6. Glossary: http://www.davidbogler.com/Pollen/glossary.html

7. Pollen wall: https://www.medicinalplantsarchive.us/pollen-grains/symmetry.html

8. PalDat – a palynological database (2000 onwards, www.paldat.org)

9. Ferguson, I. K., Webb, D. A. Pollen morphology in the genus *Saxifraga* and its taxonomic significance
   Botanical Journal of the Linnean Society, Volume 63, Issue 4, 1 October 1970, Pages 295-311
   https://academic.oup.com/botlinnean/article-abstract/63/4/295/2725773

10. Quekett Microscopical Club   http://www.quekett.org

11. http://www.oleaceae.info/

12. Wodehouse, R. P. (1935) Pollen Grains. New York & London: McGraw Hill

Index by plant name

Name	Page	Name	Page	Name	Page
Acacia dealbata	169	*Aralia japonica*	108	Blackthorn	52
Acer pseudoplatanus	223	*Arbutus* sp.	218	Bladder campion	67
Achillea millefolium	253	Archangel, yellow	254	Bluebell	53
Aconite, winter	27	*Artemesia vulgaris*	173	Borage	54
Aesculus x carnea	139	Artichoke (globe)	37	*Borago officinalis*	54
Agapanthus	28	Arum	38	Box	55
Agapanthus africanus	28	*Arum maculatum*	38	Bramble (Blackberry)	56
Ageratum	29	Ash	158	*Brassica napus*	180
Ageratum houstonianum	29	*Aster novi-belgii*	168	*Brassica oleraceae*	64
Alcea rosea	135	Balm	40	Broad leaved willowherb	246
Alder	30	Balsam, himalayan	41	Broom	57
Alkanet, green	31	Balsam, small	42	*Bryonia dioica*	58
Allium schoenoprasum	76	*Barbarea vulgaris*	249	Bryony (white)	58
Alnus glutinosa	30	Bartsia, red	43	Buckthorn	59
Alyssum, sweet	32	Bay laurel	44	Buddleia	60
Amaryllis	33	Beech	45	*Buddleja davidii*	60
Amaryllis belladonna	33	Bell heather	124	Burnet saxifrage	61
Amelanchier canadensis	212	Bellflower, creeping	93	Busy Lizzie	62
Anagallis arvensis	204	Bellflower, peach leaved	46	Buttercup	63
Annual mercury	167	*Bellis perennis*	100	*Buxus sempervirens*	55
Annual thymelaea	230	Betony	47	Cabbage, wild	64
Anthriscus sylvestris	89	*Betula pendula*	207	Cactus	65
Antirrhinum majus	210	*Bilderdykia convolvulus*	49	California lilac	70
Apple	34	Bindweed	48	*Calluna vulgaris*	126
Aquilegia	35	Bindweed, black	49	*Calystegia silvatica*	48
Aquilegia vulgaris	35	Birch, silver	207	*Camellia japonica*	66
Arabis	36	Bird's foot trefoil	50	*Campanula persicifolia*	46
Arabis albida	36	Bittersweet	51	*Campanula rapunculoides*	93
Arabis alpina	36	Black bindweed	49	*Campanula rotundifolia*	121
Arabis caucasica	36	Blackberry	56	Campion, bladder	67

Index by plant name (continued)

Plant	Page	Plant	Page	Plant	Page
Campion, white	68	*Clematis vitalba*	181	*Crocus* sp.	95
Canadian golden rod	115	Climbing hydrangea	141	Cross leaved heather	125
Candytuft	69	Clover (red)	78	*Cucurbita pepo*	164
Carpinus betulus	138	Clover (white)	79	Cupressaceae	98
Ceanothus	70	Cocksfoot	80	*Cyclamen hederifolium*	97
Ceanothus arboreus	70	*Coleus canina*	81	*Cyclamen neapolitanum*	97
Celandine, greater	71	Coltsfoot	82	*Cynara scolymus*	37
Celandine, lesser	72	Comfrey	83	Cypress	98
Centaurea cyanus	86	Common cleaver	77	*Cytisus scoparius*	57
Centranthus ruber	234	Common mallow	162	*Dactylis glomerata*	44
Cerastium arvense	109	Common toadflax	231	Daffodil	99
Chaenomeles speciosa	145	Common wintercress	249	Daisy	100
Chamaenerion angustifolium	248	Coral bells	84	Daisy, Michelmas	168
Chelidonium majus	71	Coriander	85	Dame's violet	130
Chenopodium rubrum	116	*Coriandrum sativum*	85	Dandelion	101
Cherry laurel	148	Cornflower	86	*Digitalis purpurea*	112
Cherry, flowering	73	*Cornus sanguinea*	102	*Diplotaxis tenuifolia*	239
Chickweed	74	*Corydalis lutea*	87	*Dipsacus fullonum*	224
Chicory	75	*Corylus avellana*	123	Dogwood	102
Chimonanthus praecox	251	Cosmos	88	Enchanter's Nightshade	103
Chives	76	*Cosmos bipinnatus*	88	*Endymion non-scriptus*	53
Christmas cactus	65	Cow parsley	89	*Epilobium angustifolium*	248
Cichorium intybus	75	Cowslip	90	*Epilobium hirsutum*	247
Circaea lutetiana	103	Crack willow	243	*Epilobium montanum*	246
Cirsium dissectum	225	Cranesbill	91	*Epiphyllum truncatum*	65
Cirsium erisithales	228	Cranesbill, meadow	92	*Eranthis hyemalis*	27
Cirsium palustre	226	*Crataegus monogyna*	122	*Erica cinerea*	124
Cirsium vulgare	227	Creeping bellflower	93	*Erica tetralix*	125
Cistus laurifolius	200	Creeping Jenny	94	Eucalyptus	104
Cleaver, common	77	Creeping red thyme	229	*Eucalyptus dalrympleana*	104

Index by plant name (continued)

	Page		Page		Page
Euonymus europaeus	215	*Glechoma hederacea*	143	*Hesperis*	130
Euphorbia (sun spurge)	105	Globe artichoke	37	*Hesperis matronalis*	130
Euphorbia (wood spurge)	106	Goat willow	244	*Heuchera sanguinea*	84
Euphorbia amygdaloides	106	Golden rod, Canadian	115	Hibiscus	131
Euphorbia helioscopia	105	Goosefoot, red	116	Himalayan balsam	41
Evening primrose (large)	107	Gorse	117	*Hippocrepis comosa*	140
Fagus sylvatica	45	Granny's bonnet	35	Hoary cress	132
Fallopia convolvulus	49	Grape hyacinth	118	Hogweed	133
Fatsia	108	Grass, cocksfoot	44	Holly	134
Field Mouse-ear	109	Great mullein	174	Hollyhock	135
Field scabious	203	Great willowherb	247	Honesty	136
Filipendula ulmaria	165	Greater celandine	71	Honeysuckle	137
Fireweed	248	Greater spearwort	119	Hornbeam	138
Floss flower	29	Green alkanet	31	Horse chestnut, red	139
Flowering cherry	73	Grey willow	245	Horseshoe vetch	140
Forget me not	110	Ground ivy	143	*Hyacinthoides non-scripta*	53
Forsythia	111	Groundsel	120	*Hydrangea anomala* subsp. *petiolaris*	141
Forsythia intermedia	111	Harebell	121	*Iberis gibraltarica*	69
Foxglove	112	Hawthorn	122	Ice plant	142
Fragaria × ananassa	217	Hazel	123	*Ilex aquifolium*	134
Fraxinus excelsior	39	Heather, bell	124	*Impatiens grandulifera*	41
Freesia	113	Heather, cross leaved	125	*Impatiens parviflora*	42
Fritillaria meleagris	209	Heather, ling	126	*Impatiens walleriana*	62
Galanthus nivalis	211	*Hedera helix*	144	*Ipomoea tricolor*	172
Galium aparine	77	Hedge mustard	127	Ivy	144
Galium verum	147	*Helianthus annuus*	221	Ivy leaved speedwell	213
Geranium pratense	92	Hellebore	128	*Jacobaea vulgaris*	194
Geranium pyrenaicum	91	*Helleborus* sp.	128	Japanese quince	145
Geranium robertianum	129	*Heracleum sphondylium*	133	Jasmine, winter flowering	250
Gladiolus	114	Herb Robert	129	*Jasminum nudiflorum*	250

Index by plant name (continued)

Name	Page	Name	Page	Name	Page
Juglans regia	240	London pride	156	*Mentha aquatica*	170
Knautia arvensis	203	*Lonicera periclymenum*	137	*Mercurialis annua*	167
Laburnum	146	Loosestrife (purple)	157	Michelmas daisy	168
Laburnum anagyroides	146	Lords-and-ladies	158	Mimosa	169
Lady's bedstraw	147	*Lotus corniculatus*	50	Mint (water)	170
Lamium album	178	Lucerne	159	Mistletoe	171
Lamium galeobdolon	254	*Lunaria rediviva*	136	Mock orange	186
Lamium purpureum	176	Lungwort	192	Morning glory	172
Lathyrus odoratus	222	*Lysimachia nummularia*	94	Mugwort	173
Laurel, bay	44	*Lythrum salicaria*	157	*Muscari armeniacum*	118
Laurel, cherry	148	Mahonia	160	*Myosotis scorpiodes*	110
Laurus nobilis	44	*Mahonia aquifolium*	160	*Narcissus* sp.	99
Lavandula angustifolia	149	Maize	161	Nasturtium	175
Lavender	149	Mallow, common	162	Nettle, red dead	176
Lepidium draba	132	*Malus* sp.	34	Nettle, stinging	177
Leptospermum myrtifolium	150	*Malva sylvestris*	162	Nettle, white dead	178
Leptospermum scoparium	151	Manuka	151	Nightshade, enchanter's	103
Lesser celandine	72	Marjoram	163	Oak	179
Leucojum aestivum	220	Marrow	164	*Odontites verna*	43
Ligustrum vulgare	191	Marsh thistle	226	*Oenothera erythrosepala*	107
Lilac	152	*Matricaria perforata*	205	Oil seed rape	180
Lily, arum	38	Mayweed, scentless	205	Old man's beard	181
Lime	153	Meadow buttercup	63	*Onobrychis viciifolia*	202
Limnanthes douglasii	154	Meadow cranesbill	92	*Origanum vulgare*	163
Linaria vulgaris	231	Meadow thistle	225	*Ornithogallum umbellatum*	216
Ling heather	126	Meadowsweet	165	*Oxalis acetosella*	252
Lobelia	155	*Medicago sativa*	159	*Oxalis corymbosa*	182
Lobelia erinus	155	Melilot, ribbed	166	Oxalis, pink	182
Lobularia maritima	32	*Melilotus officinalis*	166	*Papaver* sp.	190
London plane	188	*Melissa officinalis*	40	Passerina	230

Index by plant name (continued)

Name	Page	Name	Page	Name	Page
Passiflora caerulea	183	Purging buckthorn	59	Rock cress	36
Passion flower	183	Purple loosestrife	157	Rock rose	200
Peach leaved bellflower	46	Pyracantha	193	Rocket, sweet	130
Penstemon	184	*Pyracantha coccinea*	193	Rocket, wild	239
Pentaglottis sempervirens	31	*Quercus robur*	179	*Rosa rugosa*	199
Perennial wall-rocket	239	Quince, japanese	145	Rose, ramanus	199
Periwinkle	184	Ragwort	194	Rose, rock	200
Persicaria maculosa	196	Ramanus rose	199	Rosebay willowherb	248
Philadelphus	186	*Ranunculus acris*	63	Rosemary	201
Philadephus coronarius	186	*Ranunculus aquatilis*	96	*Rosmarinus officinalis*	201
Pieris japonica	187	*Ranunculus ficaria*	72	*Rubus fruticosus*	56
Pimpinella saxifraga	61	*Ranunculus lingua*	119	Sainfoin	202
Pine, scots	206	Rape, oil seed	180	*Salix caprea*	244
Pink oxalis	182	Rattle, yellow	255	*Salix cinerea*	245
Pinus sylvestris	206	Red bartsia	43	*Salix fragilis*	243
Plantago lanceolata	197	Red clover	78	*Saxifraga × urbium*	156
Platanus x hybrida	188	Red currant	195	Saxifrage, burnet	61
Plectranthus caninus	81	Red dead-nettle	176	Scabious, field	203
Plum	189	Red goosefoot	116	Scaredy cat plant	81
Polygonum persicaria	196	Red horse chestnut	139	Scarlet pimpernel	204
Poppy	190	Red valerian	234	Scentless mayweed	205
Primrose, evening	107	Redshank	196	*Schlumbergera truncata*	65
Primula veris	45	*Reseda luteola*	242	Scots pine	206
Privet	191	*Rhamnus cathartica*	59	*Scutellaria galericulata*	208
Prunus domestica	189	*Rhinanthus minor*	255	*Sedum acre*	219
Prunus laurocerasus	148	Rhododendron	198	*Sedum maximum*	142
Prunus sp.	73	*Rhododendron ferrugineum*	198	*Senecio jacobaea*	194
Prunus spinosa	52	Ribbed melilot	166	*Senecio vulgaris*	120
Pulmonaria	192	*Ribes rubrum*	195	*Silene alba*	68
Pulmonaria angustifolia	192	Ribwort plantain	197	*Silene vulgaris*	67

Index by plant name (continued)

Plant	Page	Plant	Page	Plant	Page
Silver birch	207	Sweet alyssum	32	*Ulex europaeus*	117
Sisymbrium officinale	127	Sweet pea	222	*Urtica dioica*	177
Skullcap	208	Sweet rocket	130	Valerian, red	234
Small balsam	42	Sweet violet	238	*Verbascum thapsus*	174
Snake's head fritillary	209	Sycamore	223	Verbena	235
Snapdragon	210	*Symphyotrichum novi-belgii*	168	*Verbena x hybrida*	235
Snowdrop	211	*Symphytum officinale*	83	*Veronica chamaedrys*	214
Snowy mespilus	212	*Syringa vulgaris*	152	*Veronica hederifolia*	213
Solanum dulcamara	51	*Taraxacum* sp.	101	Vetch tufted	236
Solanum lycopersicum	232	*Taxus baccata*	256	Vetch, horseshoe	140
Solidago canadensis	115	Teasel	224	Viburnum	237
Sorrel, wood	252	Thistle, marsh	226	*Viburnum lantana*	241
Spear thistle	227	Thistle, meadow	225	*Viburnum tinus*	237
Spearwort, greater	119	Thistle, spear	227	*Vicia cracca*	236
Speedwell (germander)	214	Thistle, yellow melancholy	228	*Vinca minor*	185
Speedwell, ivy leaved	213	Thyme, creeping red	229	*Viola odorata*	238
Spindle tree	215	*Thymelaea passerina*	230	Violet, sweet	238
Spring crocus	95	*Thymus* sp.	229	*Viscum album*	171
Spurge, sun	105	*Tilia* sp.	153	Walnut	240
Spurge, wood	106	Toadflax, common	231	Water crowfoot	96
Stachys officinalis	47	Tomato	232	Water mint	170
Star of Bethlehem	216	Trefoil, bird's foot	50	Wayfaring tree	241
Stellaria media	74	*Trifolium pratense*	78	Weld	242
Stinging nettle	177	*Trifolium repens*	79	White bryony	58
Stonecrop	219	*Tripleurospermum inodorum*	205	White campion	68
Strawberry	217	*Tropaeolum majus*	175	White clover	79
Strawberry tree	218	Tufted vetch	236	White dead-nettle	178
Summer snowflake	220	Tulip	233	Whitetop	132
Sun spurge	105	*Tulipa* sp.	233	Wild cabbage	64
Sunflower	221	*Tussilago farfara*	82	Willowherb, broad leaved	246

Index by plant name *(continued)*

	Page		Page		Page
Willowherb, rosebay	248	Wintercress, common	249	Yellow corydalis	87
Willow, crack	243	Wintersweet	251	Yellow melancholy thistle	228
Willow, goat	244	Wood sorrel	252	Yellow rattle	255
Willow, grey	245	Wood spurge	106	Yew	256
Winter aconite	27	Yarrow	253	*Zantedeschia aethiopica*	38
Winter flowering jasmine	250	Yellow archangel	254	*Zea mays*	161